Life's Origin

Life's Origin

The Beginnings of Biological Evolution

Edited by

J. William Schopf

UNIVERSITY OF CALIFORNIA PRESS
Berkeley · Los Angeles · London

University of California Press
Berkeley and Los Angeles, California

University of California Press, Ltd.
London, England

© 2002 by the Regents of the University of California

Library of Congress Cataloging-in-Publication Data

Schopf, J. William, 1941—.
  Life's origin : the beginnings of biological
evolution / edited by J. William Schopf.
      p.   cm.
  Includes bibliographical references and index.
  ISBN 0-520-23390-5 (alk. paper).—ISBN 0-520-
  23391-3 (pbk. : alk. paper)
  1. Life—Origin   I. Title.

QH325.L694 2002
576.8'3—dc21   2002002071

Manufactured in the United States of America
10   09   08   07   06   05   04   03   02   01
10   9   8   7   6   5   4   3   2   1

The paper used in this publication meets the minimum requirements of ANSI/NISO Z39.48-1992
(R 1997) (*Permanence of Paper*).

Contents

The What, When, and How of Life's Beginnings

J. WILLIAM SCHOPF

In this news-conscious age, everyone knows the journalist's litany of prime questions: who, what, when, where, why, how. About the origin of life, scientists' questions are similar, but more restricted. Setting aside the *who* (no humans were on hand to observe the event), the *where* (unanswerable except in the broadest terms—on Earth, in water, probably in oceans), and the *why* (a question posed by philosophers and theologians, not scientists), we are left with the three great puzzles this volume addresses: *What* is the origin of life, *when* did it begin, and *how*?

At first blush, the *what* seems easy to answer: Surely, we can posit a primal protoplasmic globule, a first speck of living matter. But the answer is not so simple. Just as the colors of a rainbow merge imperceptibly one into another, the transition to life from nonlife is a seamless continuum. Some researchers define life as an interacting combination of a particular set of chemicals. Others consider it a highly ordered, intricately complex information-processing system. Still others insist upon a full-blown living cell. Consensus is elusive. Although the extremes of the continuum are sharp and clear, the dividing line between nonlife and life is blurred.

Surprisingly, deciphering the *when* of life's beginnings is even more vexing. Life's origin ought to be easy to date—trace the fossil record back to a time when the record peters out, and, voilà, that's when life began. But Nature has proven uncooperative, for it is the rock record that peters out, not the evidence of life. Over time, Earth's rock record has been weathered away and lost by the geologic cycle of uplift, erosion, burial, and pressure-cooking (followed by more uplift, erosion, burial, and cooking); thus, few traces of the planet's early history are left today. However, of the slivers of ancient terrains that do remain, most contain traces of life. This is important knowledge. By telling us that life started remarkably early in the history of the planet, it suggests that life's origin happened more quickly—and much more easily—than we otherwise might have imagined. But it has a downside. Hidden deep in the recesses of Earth's past, dating from a time for which no surviving rocks can tell the story, hard evidence not only of life's beginning but the life-generating event itself seems lost forever.

Of the three great puzzles, the *how* of life's beginnings has been plumbed most deeply. But here, too, the answer is hazy. (Were it not, the names of the Nobel Prize–winning researchers would be familiar to us all and this book would be quaintly dated.) But given the depth of this problem and the role it plays in understanding where living systems (humans included) fit in the natural order, it is not surprising that this great mystery is as yet unsolved.

Until fairly recently, the origin of life was sacrosanct territory, the exclusive province of theologians. These limits were breached in 1953 when Stanley Miller announced a breakthrough discovery. He showed that amino acids like those present in living systems can be made easily, in a matter of days, from very simple ingredients, in the total absence of life. Carried out under conditions designed to mimic those of the primitive Earth, his experiments were the first to show a plausible way that potentially life-generating organic molecules could have formed before life got started. Yet even with this impetus, the subject remained taboo in some quarters. For example, the late Sidney Fox, a pioneer in origin-of-life studies, liked to remind his colleagues that in the mid-1950s, the fledgling National Science Foundation refused his request for funding—not on scientific grounds, but to avoid being torn asunder by fundamentalists in Congress. Since that time, the origin of life has gained its rightful place among the great unsolved questions of science. Ironically, however, the flood of new research has pointed up just how complicated the problem is. This, too, is unsurprising. In science, the devil *is* often in the details: Time and again, simple, first-found solutions turn to dust as further work uncovers new layers of unexpected complexity.

Savants have long pondered what life is and how it differs from inanimate matter. How did the ancients explain its origin? How and by whom did modern studies get started? Where do living systems fit in the cosmic panorama? And if life exists beyond the Earth, how can it be discovered? These are the subjects of chapter 1, an introduction to the volume that spans a breadth of issues addressed in greater depth in later chapters. But this essay is more than an overview. Written in a highly personal vein that shows a human side of science rarely revealed in the scientific literature, it should remind us that the search for life's beginnings is an enterprise carried out by researchers whose manifest zeal reflects deep-seated human feelings—scientists who are themselves more than a little in awe of the prize they seek, an understanding of their own deep roots.

Where does the problem of the origin of life stand today? As this volume shows, quite a lot is known, at least in broad outline. So even though detailed understanding has yet to be achieved, the main story of life's beginnings is abundantly clear: Life is a natural outcome of the evolution of cosmic matter. Thus, the narrative begins not on Earth, but in the interiors of distant stars, not with molecules of living cells, but with the nucleosynthesis billions of years ago of a particular small cluster of life-generating (biogenic) chemical elements—carbon, hydrogen, oxygen, nitrogen, sulfur, and phosphorus (CHONSP)—elements that combined to form the biochemicals of every earthly living system. This story, from the genesis of the biogenic elements to their bonding into bodies of the Solar System, Earth included, is the focus of chapter 2, a magnificent tour through the evolution of the cosmos from the Big Bang to the formation of our planet.

With the stage set for the origin of life—on a lifeless, wave-washed, volcanic rocky planet—chapter 3 shows how the first of three critical steps toward life's beginnings may have taken place. Not only is all life mostly composed of atoms of CHONSP, but in every living system these elements always make up the same special suite of building-block molecules, or monomers—amino acids, sugars, purines, pyrimidines, and the like. So, just as life could not exist without CHONSP, life without these monomers (or compounds similar enough to them to take their place) is all but inconceivable. Laboratory work shows that these crucial building blocks could easily have formed in a variety of settings on the primordial Earth. While some laboratory protocols are more efficient than others and some more closely represent the lifeless primitive environment, all give products similar to biologic molecules if (as on the early Earth) free oxygen, a poison to such systems, is in scant supply. Of all parts of the origin-of-life scenario, this first step, the nonbiologic buildup in Earth's early oceans of chemically simple organic monomers, is best understood.

But life, of course, is far more than a collection of simple monomers. Almost all of life's key compounds are polymers, large molecules such as carbohydrates, proteins, and gene-carrying DNA. Though such polymers can be huge, containing millions of atoms, they are often surprisingly uncomplicated—slender threadlike strands made up of regular sets of simple subunits strung together like colored beads on meandering strings. Here, again, the relative simplicity and chemical sameness of all life points the way in origin-of-life research. Not only is all life mostly CHONSP, always present in the same small suite of building-block monomers, but the same kinds of monomers are linked together in all organisms to form the same few kinds of polymers. The step from monomeric building blocks to the polymers of life, the second of the three critical steps in life's beginnings and the subject of chapter 4, is thus among the most important in unraveling how life got started. But it is also one of the most challenging. Cells build polymers by adding monomers, one at a time, to the end of a growing polymeric chain, a process speeded by enzymes and powered by cellular energy that gives off a molecule of water every time a monomer is added. But neither cellular energy nor complicated enzymes were present on the lifeless planet. And since polymer formation requires the removal of water from the growing chain, the buildup of polymers in the watery milieu of the early ocean would have been quite a feat— akin to drying one's hands on a sopping washcloth. Still, several prebiotic ways have been devised to turn the trick. Among these, the most ingenious and perhaps likely is the assembly of such polymers on particles of clay like those sedimented at the margins of shallow lagoons. Though this part of the puzzle is not as clearly defined as monomer formation, a plausible pathway to polymer formation seems reasonably well worked out.

Chapter 5 addresses the third major step in the life-generating scenario, the nonbiologic formation of gene-like, information-containing molecules. Though arguably the most crucial step, it is only partly understood. Origin-of-life research is driven by the chemical sameness of all life a uniformity that extends to the organization of cells. And just as all cells are made up of nearly identical suites of chemical elements, monomers, and polymers, sameness also characterizes the way they work. All are based on but a single kind of informational gene-carrying polymer, the nucleic acid DNA. The life-directing message encoded chemically in DNA is transferred to another kind of nucleic acid, RNA, which carries the information to tiny bodies in the cell, where it is used to make proteins such as enzymes. This intricate system is much too complicated to date from life's beginnings; moreover, DNA itself is advanced, the evolutionary descendant of earlier evolved polymers of RNA. So the present-day DNA–RNA–protein system of life seems likely to have been rooted in an earlier system of organization—that of the "RNA World," in which molecules of RNA served as the sole carriers of genetic information. But herein lies the rub: RNA could not have been plentiful on the lifeless Earth unless the sugar ribose, one of its prime components, was available in copious supply. And, as laboratory experiments show, such a sugar abundance is highly unlikely. So the RNA-based system must itself have been derived from even earlier forms of life that, lacking both RNA and DNA, would have been decidedly different from anything alive today. Therefore, the biochemistry of living organisms can be expected to provide only the most limited clues to how life began. In short, though a great deal has been learned about the deep history of information-containing gene-like systems, the origin of the RNA World is a subject ripe for further research.

Taken together, the first five chapters present a plausible scenario of how life got started. But they tell us little about what life forms emerged first and when such life began. These key questions are addressed in chapter 6. Three lines of evidence are brought to bear. First, the biology of primitive microbes living today is used to identify traits that date from near life's beginnings, hints of what early life may have been like. Second, life's roots are traced back through the ancient fossil record of tiny petrified cells and biochemical imprints left in rocks to show what kinds of organisms were present early and how long life has existed. Third, evidence from the formative stages of the planet's history is used to shed light on when Earth's early hostile environment first became clement enough to harbor living systems. Collectively, the evidence reveals that microbial life was diverse, flourishing, and presumably widespread as early as 3,500 million years ago—only a few hundred million years after settings had become suitable for microbes to gain a foothold—and that life on Earth evolved far and fast in a remarkably short time.

As the essays in this volume show, an overall picture of how life emerged is achieving ever-clearer focus. But the search for deep knowledge of the processes involved, an understanding of the details of each step, is a work in progress. With active research dating only from the 1950s, the field is young, vibrant, and advancing rapidly. Given time, effort, and a continuing influx of imaginative students and fresh ideas, we can one day fully answer the *what, when,* and *how* of life's beginnings.

CHAPTER I

Historical Understanding of Life's Beginnings

JOHN ORÓ

WHAT IS LIFE?

There are three major singularities in the world—the observable universe, life on Earth, and human beings. For the most part, we agree on concepts of what the cosmos and human beings are, but we have reached no consensus on what life is. It is easier to recognize life, in all its common forms, than it is to define it. In 1976, during NASA's Viking Project to Mars, the late Carl Sagan and I were at the Jet Propulsion Laboratory in Pasadena, California, discussing this very problem. Asked if we would actually recognize life on Mars were we lucky enough to encounter it, Carl answered with ebullient humor: "John! If a herd of elephants stampeded across the field in front of the camera, we would not have any doubt about the existence of life on the Red Planet. Is that simple enough?"

Well, perhaps. But barring stampeding elephants, probably not. Still, life can be defined in relatively simple terms, as many encyclopedias have done. A common definition, for example, is that "Life is a dynamic state of organized matter characterized basically by its capacity for adaptation and evolution in response to changes in the environment, and its capacity for reproduction to give rise to new life. Such a state is a consequence of metabolic reactions (**anabolism** and **catabolism**) and of the interaction of the living organism with other organisms and the environment." More generally, however, we describe the nature of life in terms of the main attributes that characterize living systems, as follows:

1. Dynamic, self-organized, independent structural entities

2. Made of water (H_2O), the biogenic elements **CHONSP** (carbon, hydrogen, oxygen, nitrogen, sulfur, and phosphorus), and organic molecules

3. Able to extract and recycle matter and energy from the environment

4. Self-reproducing by translation of stereospecific informational polymers

5. Adaptable and subject to mutation and evolution by natural selection

6. Metabolize by self-regulated stereospecific catalytic reactions

7. May exhibit either unicellular organization or multicellular organization and differentiation

8. Exhibit organismic growth and asexual or sexual reproduction

9. Dynamically interact with and sensorially respond to stimuli

10. May exhibit conscious thought, language, writing, and technology.

The basic nature of life may seem obvious, rendering a formal definition unnecessary or, at most, secondary in importance. However, a definition of life's attributes is pivotal to at least three areas of science—studies of the origin of life, **artificial life**, and **exobiology**. Thus, from a physicochemical point of view, A. Moreno (1998) has offered a definition of a living system in tune with the principles of irreversible **thermodynamics**: "A type of dissipative chemical structure which builds recursively its own molecular structure and internal constraints and manages the fluxes of energy and matter that keep it functioning (by **metabolism**), thanks to some macromolecular informational registers (DNA, RNA) autonomously interpreted which are generated in a collective and historical process (evolution)." In other words, we can say that a living system is a self-maintained organic structure operating in an aqueous medium as a self-regulated **negentropic process** that exchanges matter and energy with the environment and is capable of self-reproduction, evolution by natural selection, and adaptation to the environment.

Yet another definition has been offered by C. Grobstein (1965): "Life is macromolecular, hierarchically organized, and characterized by replication, metabolic turnover, and exquisite regulation of a spreading center of order in a less ordered universe." But perhaps the shortest definition of life is that proposed by L.E. Orgel (1973): "Living organisms are **CITROENS**—Complex Information-Transforming Reproducing Objects that Evolve by Natural Selection."

Alternatively, we could approach the definition of life by asking a simple question: How would one describe the many different kinds of life? In general, we could answer that there are at least ten different kinds of life, each with its own definition:

1. Intelligent human life
2. Animal life (all kinds)
3. Plant life (all kinds)
4. Fungal life (all kinds)
5. Eukaryotic multicellular life
6. Eukaryotic unicellular life
7. Prokaryotic commensal and interdependent cellular life
8. Prokaryotic unicellular life
9. Complex and simple viruses
10. Subcellular and molecular life (self-replicating and autocatalytic, having DNA, RNA, or functionally equivalent informational macromolecules)

This list is not exhaustive, but it does implicitly recognize all possible variations of symbiotic organisms (such as lichens) and other life forms (for example, micoplasma). However, when we get to the realm of viruses and subcellular and molecular life, the task of defining life becomes exceedingly tricky. When chemists look at viruses, they see a complex organization of molecules. But when biochemists look at a virus, they see something more. In addition to its complexity of nucleic acids, **DNA** (deoxyribonucleic acid) or **RNA** (ribonucleic acid), and a few key proteins, they see an entity resembling a **protobiont,** a primitive precellular living system, perhaps not greatly different from life forms that existed on Earth 4,000 million years ago. This view of the similarity between viruses and protobionts may not be generally accepted. But it seems to me that when talking about life of the precellular **RNA World,** we are really talking about a level of organization like that of a virus. (As an aside, one of my students, Richard Carlin, has measured the decreasing length of the polyadenylic tail of various RNA viruses and plotted the age of the host they infect. Extrapolating, he has estimated that life originated on Earth between 3,800 and 4,000 million years ago, a date consistent with most current estimates.)

Concepts about the Nature of Life

Concepts and statements about the nature of life date back to antiquity. Consider the Greek philosopher Democritus (~460 B.C.E.), who said, "Life is the result of a special combination of atoms characterized by their constant mechanical movement." In our own era, the Russian biochemist A. I. Oparin said, "Life is a qualitatively peculiar form of organization and movement of matter, which came about as a result of a series of natural processes during a certain stage in the history of the Earth as part of a developmental process of the evolution of matter" (Oparin 1986). And in his book *What Is Life?* the Nobel Prize–winning Austrian physicist Erwin Schrödinger suggested that the structure of life must be that of an aperiodic (nonuniform) crystal (Schrödinger 1956). By Schrödinger's time, it had become apparent that, in fact, "life is chemistry" and that the basic material of heredity is the helical, ladder-like compound DNA. The discovery of DNA showed that some biological molecules are capable of making copies of themselves. In reality, the aperiodic repetitive crystal structure suggested by Schrödinger is the same as that modeled by James Watson and Francis Crick—a crystal of the molecule of DNA, the "essence of life" (Margulis and Sagan 1997). Of course, definitions of life are sometimes of a lighter vein. One Hindu proverb, for instance, has it that "Life is a bridge—cross it, but don't build a house on it."

Is All Life Based on Carbon?

To date, we have but one example of life—life on Earth. So we cannot prove that life must be based on carbon. In principle, we are in no position to judge because we have only the statistics of a single example. However, if we reflect on all we know about life, it is very difficult to imagine the existence of any kind of life that is not based on the materials of which we ourselves are made—water (H_2O) and the biogenic elements (CHONSP). Once I was giving a lecture on the origin of life and the possibility of life beyond the Solar System when Carl

Sagan spoke up, interrupting my presentation: "But John! Could we not have a life based on tungsten, for instance?" I answered, "Carl, I don't think so, but if that were the case, the ball is in your court. Please, show me!" For a short while, I let the answering silence speak for itself; then I had to explain the inherent impossibility of life based on tungsten, a heavy metal: Chemically, no heavy metal is capable of performing the functions that carbon serves. (Equally improbable is any kind of life existing in a vacuum, as Freeman Dyson proposed at a recent meeting of the International Society for the Study of the Origin of Life, or of a "living clay," as Graham Cairns-Smith has conjectured.)

I am persuaded that the secret of this mystery, this singularity we call life, resides fundamentally in organic chemistry—that is, in the capacity of organic structures to react and interact in a self-organized manner. And organic chemistry is based, in turn, on the prime property of the element carbon, which can combine not only with itself but with all the other biogenic elements to generate molecular structures of great diversity. It is this profusion of combinations, catalytic activities, and interactions that operates in the aqueous and lipid-rich environment of cells.

The fundamentals of organic chemistry give life an upside and a downside, building cells from carbon compounds and breaking down biochemicals when an organism dies. And both sides are important. Indeed, the fact that Earth's ecology continuously recycles the constituent elements of life is critical. In the energy cycle of life, carbon dioxide (CO_2) is fixed in a plant by the light-powered process of photosynthesis (light energy $+ CO_2 + H_2O \rightarrow CH_2O + O_2$); then the plant, with its organic nutrients, is eaten by an animal. Taking oxygen (O_2) from the atmosphere, the animal oxidizes the nutritious organic matter to obtain energy ($O_2 + CH_2O \rightarrow CO_2 + H_2O$ + energy), then returns CO_2 to the atmosphere. If life were silicon-based, instead of giving off gaseous CO_2 into the environment, the process would yield crystalline silicon dioxide (SiO_2)—sand. Being a solid, SiO_2 would simply accumulate, thereby prohibiting a biogeochemical cycle! (The same would be true were tungsten to be substituted for carbon. Such differences apply, of course, only to conditions at the surface of the Earth, not necessarily to those of much hotter settings, such as the cores of stars, where SiO_2 would be a gas.)

In short, it is difficult to conceive of a kind of life that is not based primarily on the six biogenic elements—the prime elements that make up the **nucleic acids, sugars, proteins, phospholipids,** and all the **enzymes** and **coenzymes** on which Earthlife depends. To these critical elements, we should add various metallic ions (for example, magnesium, iron, and zinc), which are central to the structures of **chlorophyll, cytochromes, hemoglobin,** and numerous enzymes. But life as we know it is CHONSP. And, as far as we now know, CHONSP-life is the only life there is.

BELIEFS AND HYPOTHESES ON THE ORIGIN OF LIFE

Materialism and Idealism

Quite a number of ancient writings address the origin and nature of life from a materialistic or idealistic point of view. The Indian vedas and upanishads, for example, consider as a fact the existence of particles of life that pervade the universe. Similar concepts were developed in greater detail by the ancient Greeks (particularly, Thales, Anaxagoras, Democritus, Epicurus, and their followers). The best known of the Greek philosophers (at least in today's world) are the Socratics—Plato (427?–347 B.C.E.) and his pupil Aristotle (384–322 B.C.E.)—whose thinking was suffused with idealism. For Plato, a pure idealist, everything in the natural world derived from nonmaterial eternal ideas. Aristotle modified Plato's ideas to embrace a form of dualism. He argued that the material world has its own existence and should be intensively described. But he also maintained that living beings, in addition to having material substance, are endowed with a vital principle, an **entelechy** (from the Greek *entelecheia*), that regulates and directs the vital processes of each organism. This entelechy is nonmaterial and therefore undiscoverable by scientific investigation. In contrast, the pre-Socratic Thales (640?–546 B.C.E.) thought that water, the most pervasive element of matter, was the origin of all things, including life. (Indeed, the impressive power of the waters, from the life-giving Nile to the oceans that envelop the world, still shapes human thinking. Consider, for example, the inhabitants of Papua New Guinea's Trobriand islands, who believe that human life is conceived through spirits that float in the ocean when young women and men bathe in the South Pacific waters.)

Creationism

After the conquests of Alexander the Great (356–323 B.C.E.), who was the pupil of Aristotle, the philosophic traditions of ancient Greece met and clashed with a different cosmogenic idea from the Near East. The Semitic People of the Book found in their sacred scriptures a supernatural source of all life on Earth, including humans. For the adherents first of Judaism, then Christianity and Islam, the world is a creation, the handiwork of a single Creator. When, through the agency of Rome, the Judeo-Christian tradition (seasoned with Greek idealism) spread to Europe in the Common Era, its cosmogeny informed the thinking of all Western natural philosophers. Indeed, the mystery of life is so great, and the biblical tradition so powerful, that creationism (in its diverse forms) continues into the modern era. Newton (1642–1727), who originated present-day physics and invented the calculus, spent many years trying to prove the existence of God. And in our own century, many prominent scientists continue to bow, sometimes gingerly, before a Creator. Erwin Schrödinger remarked that "life is the work of the fine and precise creation of the quantum mechanics of our Lord" (see Oparin 1986), and Sir John Eccles has written that "evolutionary Darwinism culminates by the infusion in the brain of *Homo sapiens sapiens* of the soul by our Creator" (Eccles 1991).

Spontaneous Generation

From ancient times almost to the present century, the belief that many organisms could originate from inert matter was common. Early Egyptians, for instance, thought that life arose from the muds of the Nile. This belief was widespread throughout the Renaissance. For instance, the Belgian physician and chemist Jan Baptista van Helmont (1580–1644) provided a detailed recipe for generating mice in 21 days from open jars stuffed with soiled underwear and kernels of wheat.

Spontaneous generation, the idea that life can originate all at once from dead matter, remained entrenched until the nineteenth century, when Louis Pasteur (1822–1895) made his discoveries. In addition to studying the process of fermentation, discovering the germs that cause many diseases, developing vaccines, and inventing pasteurization (the method of sterilization named in his honor), Pasteur investigated spontaneous generation experimentally. Studying various solutions heated in swan-necked flasks, he firmly refuted the theory. In Pasteur's view, living systems could arise only from other living systems: *Omni vivum ex vivum, omni ovo ex ovo.*

Cosmic Panspermia and the Life Cloud

The existence throughout Nature of a fundamental "carrier of life" was expressed in the earliest Hindu sacred writings, the vedas, and formulated more fully by the Greek philosopher Anaxagoras (500?–428 B.C.E.). Anaxagoras embraced the concept of **panspermia**, the idea that life exists everywhere. During the nineteenth and early twentieth centuries, several investigators, most notably the Swedish physicist and chemist Svante August Arrhenius (1859–1927), proposed that, given Pasteur's refutation of spontaneous generation, life must have originated extraterrestrially. Arrhenius suggested that just as organic matter could be transported on meteorites coursing through the Solar System (as first shown in the 1830s by Berzelius's analysis of organic matter in the Aläis **carbonaceous meteorite**), the germs of life could be transported between the stars by the radiation pressure of starlight.

A more recent version of this hypothesis was proposed by Francis Crick and Leslie Orgel. They have suggested the possibility of "directed" panspermia—the introduction of life to Earth by intelligent beings, implanted either directly or by robotic means. The latter, in fact, may already have happened, at least from Earth to other bodies of the Solar System. Microbes carried on our automated spacecraft may have contaminated other planets, such as Mars. And the transport of small rocky fragments from one body of our Solar System to another is a reality, as meteorites have come to Earth from the Moon and Mars. What has not yet been demonstrated is an occurrence of interplanetary panspermia, the delivery of a viable life form from one planetary body to another. Furthermore, the theory of cosmic panspermia offers no solution to the origin of life: It simply relegates that process to an undefined setting in some unknown place at some indeterminant time elsewhere in the cosmos.

An extreme expression of the concept of cosmic panspermia is that proposed by Fred Hoyle and Chandra Wickramasinghe. Seizing on the similarity between the infrared spectrum of **interstellar dust** and that measured on preparations of the common human gut bacterium *Escherichia coli*, they argue that interstellar dust is a "life cloud" composed largely of bacteria. Some years ago, during a luncheon at Rice University in Houston, I tried to explain the impossibility of this hypothesis to Fred Hoyle. I pointed out that the spectral similarity is only partial. Moreover, such a degree of similarity is exactly what we should expect, given the bacterium-like suite of biogenic elements present in interstellar dust (shown in table 1.1), an elemental composition shown by several investigators, most recently by Delsemme (2000).

In 1981, I took part in a debate on comets and the origin of life at the University of Maryland as part of a scientific meeting organized by

TABLE 1.1 PERCENTAGES OF ELEMENTS IN LIFE
COMPARED WITH THOSE IN INTERSTELLAR FROST
AND COMETS

Elements	Bacteria	Mammals	Interstellar Frost	Volatiles in Comets
Hydrogen	63.1	61.0	55	56
Oxygen	29.0	26.0	30	31
Carbon	6.4	10.5	13	10
Nitrogen	1.4	2.4	1	2.7
Sulfur	0.06	0.13	0.8	0.3
Phosphorus	0.12	0.13	—	0.08
Calcium	—	0.23	—	—

Adapted from Delsemme (2000).

the late Cyril Ponnamperuma (1981). I asserted that the life cloud idea presented earlier in the meeting by Dr. Wickramasinghe was "utter nonsense." A few days later, my remarks were reported in the *New York Times* by a science writer present at the meeting. Now, some two decades later, my opinion remains unchanged. Moreover, even were it to prove correct, the general hypothesis of panspermia, as promoted by Arrhenius, Hoyle, Wickramasinghe, and others, would in no way account for the origin of life. Rather, it would relocate the problem to another time and place in the cosmos but beyond the reach of experimental inquiry and serious scientific investigation.

CHEMICAL EVOLUTION

A. I. Oparin, J. B. S. Haldane, and J. D. Bernal

In 1967, during the Seventh International Congress of Biochemistry in Tokyo, I asked Alexandr Ivanovich Oparin how he had arrived at the idea of chemical evolution as an explanation for the origin of life. He replied that the basis of the idea came from Dimitry Mendeleev (1834–1907), the Russian chemist who devised the periodic table. According to Mendeleev, the petroleum in Earth's crust was generated when superheated water from the interior of the planet passed through geological strata that contained **iron carbides** (carbon–metal compounds). The reaction with these carbides produced the **hydrocarbons** (organic compounds composed solely of atoms of hydrogen and carbon) of which petroleum is composed. Professor Oparin said that when he became interested in the problem in his student days, he simply took Mendeleev's idea a step further. He hypothesized that, on the primitive Earth, the reaction of water with carbide and similar chemical reactions formed many kinds of organic compounds from which life arose through a long process of chemical evolution. This, he thought, might provide the "missing first chapter" to Darwin's great book on the evolutionary history of life.

We know today that petroleum is a result of the decay of living matter, not a product of reaction between hot percolating waters and iron carbides. But it is interesting that, through serendipity, Oparin was able to develop a reasonable hypothesis from a partially incorrect one. Partially, because, even though petroleum is unquestionably the decay product of living systems, Mendeleev's idea was not far off the mark. In my laboratory, we have produced small yields of deuterated low-molecular-weight aliphatic hydrocarbons by reacting iron carbides with deuterium-tagged water or deuterium chloride.

New discoveries notwithstanding, it was the synergistic interaction of concepts proposed by two particularly creative scientists—Mendeleev in chemistry and Charles Darwin (1809–1882) in biology—that led Oparin to his idea of how life began. Oparin's final scientific paper (Oparin 1986), published six years after his death, critically reviews various hypotheses of the origin of life. And his words still resonate:

> We can conclude that the different forms of life are not the result of a process having a determined finality developed a priori by a creative plan, nor are they the result of a chance fortuitous act. Life emerged as the result of natural evolutionary processes, as a new form of movement of matter during its process of development. The study of this process allows us to know, with a scientific base, the essence of life and its qualitative difference from the world of inorganic matter.

In the late 1920s, a hypothesis very similar to Oparin's was developed independently by the English biologist J. B. S. Haldane. He and Oparin proposed that the origin of life on Earth was preceded by a period of nonbiological molecular evolution. The Oparin-Haldane hypothesis assumes that, during and after Earth's formation, simple carbon compounds were transformed into increasingly more complex ones by means of heat and solar radiation. In addition, Oparin proposed that, after the production of organic compounds of fairly high molecular weight, a phase separation occurred, resulting in formation of microscopic organic droplets (colloidal **coacervates**). These coacervates were able to selectively accumulate simple organic compounds from the environment, thereby giving rise to chemically coupled reactions of increasing complexity—reactions that were, in important respects, similar to those of living systems. This phase separation made primitive competition possible and established a type of natural selection among pre-life, protobiont droplets. From such competition, an ancestral cell able to undergo Darwinian evolution would eventually emerge, marking the start of true biological evolution.

Darwin traveled the world, seeing it with his own eyes. As the naturalist aboard H.M.S. *Beagle,* he made an extraordinary number of observations that allowed him to compare various living species. Oparin traveled little, but read much. Absorbing multiple philosophical and scientific publications, he sought to fit pertinent scientific discoveries into his central idea of prebiological chemical molecular evolution, itself the offspring of a marriage of Mendeleev's and Darwin's theories. Further insight into the development of Oparin's ideas can be found in two excellent reviews, one by Oparin himself (1986), the other by Antonio Lazcano (1992).

Another very similar hypothesis about prebiotic chemical evolution was presented in the late 1940s by Professor J. D. Bernal of Birbeck College, London. Known for his work in crystallography and his writings on the social relevance of science—notably, *The Social Function of Science* (Bernal 1939) and *World without War* (Bernal 1958)—Bernal was also the chairman of the World Council for Peace. In 1949, he published an influential paper, "The Physical Basis of Life" (Bernal 1949), which he expanded into a book of the same title (Bernal 1951). Later he summarized his ideas in another book, *The Origin of Life* (Bernal 1967). This book contains several appendixes, including a translation of Oparin's 1924 *Proiskhozhedenie Zhizni* and a reprinting of Haldane's seminal 1929 publication. Bernal's own contributions emphasized the properties of clays as potential catalysts in chemical reactions (discussed in chapter 4) and the role of organic lipids in phase separation and the formation of cell-enclosing membranes. Unlike many others at the time, he also considered primitive Earth's capture of organic molecules from comets and meteorites important.

The First Experiments

In 1936, Oparin published a book that expanded on his earlier views. When Sergius Morgulis translated this book into English (Oparin 1938),

Oparin's views became better disseminated but did not gain the international scientific community's full acceptance. Among those who dismissed the relevance of nonbiological syntheses to the origin of life was Melvin Calvin of the University of California, Berkeley. Together with several colleagues, he published the results of a preliminary experiment on the irradiation of a "primitive atmosphere" by high-energy helium ions (Garrison et al. 1951). Calvin was a biochemist whose work focused on the fixation of CO_2 by photosynthesis in plants (work for which he received a Nobel Prize). Therefore, he did not follow Oparin's proposal of a highly reducing atmosphere of **methane** (CH_4), hydrogen (H_2), water vapor ($H_2O\uparrow$), and **ammonia** (NH_3). Rather, he used a mixture of CO_2, H_2, and $H_2O\uparrow$. Because no nitrogen was present, no **amino acids** (characterized by the amino group, NH_2) were formed (Calvin 1969). The experiment yielded small amounts of **formaldehyde** (CH_2O) and **formic acid** (HCOOH).

Another major scientist who became interested in this problem was Harold Clayton Urey, a cosmochemist (and also a Nobel laureate) at the University of Chicago. He was much involved in studies of Earth's primitive atmosphere. By 1952, he had come to Oparin's conclusion—that Earth's primitive atmosphere must have been highly reducing, having carbon in the form of CH_4 instead of CO_2 (Urey 1952). Obviously, the Calvin group's use of CO_2 instead of CH_4 opened them to criticism, as Urey suggested in a lecture to the University of Chicago's chemistry department. Among the attendees was a young Stanley L. Miller, at that time a graduate student who had just completed his undergraduate work at Berkeley.

Although Urey advised Miller to do his doctoral research on problems of inorganic cosmochemistry, Miller persuaded Urey to let him work on the problem of organic synthesis on the primitive Earth instead. In 1953 Miller carried out a now-classic experiment—he subjected a reducing "primitive atmosphere" composed of CH_4, H_2, $H_2O\uparrow$, and NH_3 to the action of a continuous electric discharge for a period of one week (Miller 1953, 1957).

Miller's experiment showed for the first time that amino acids can be produced under simulated "primitive Earth conditions" and that a number of these amino acids are of the kinds present in biological proteins. Perhaps more significant, this and subsequent experiments showed that many of the amino acids formed are comparable, both qualitatively and quantitatively, to amino acids present in the **Murchison meteorite,** a carbonaceous "shooting star" that fell to Earth in 1969 near Murchison,

Australia. This striking similarity suggests that prebiotic processes similar to those simulated in experiments such as Miller's may have occurred on the parent body from which the Murchison meteorite originated. As expected, the amino acids formed in Miller's experiments were present as a racemic mixture. And in this mixture, they occur both in the configuration in which they are present in proteins and in the nonbiological mirror image of this configuration. It is possible that they differ slightly in this regard from the Murchison amino acids, since recent studies (Cronin and Pizzarello 1997; Pizzarello and Cronin 1998) show that several amino acids in the meteorite vary by small amounts from pure racemic mixtures. This observation requires further investigation. But whether the Murchison amino acids prove to be completely or partially racemic, the suite of amino acids present in the meteorite bears marked similarities to that synthesized experimentally.

Miller's 1953 experiment stands out not only in its primacy but in its results. Because many of the types of amino acids produced are present in proteins, they are of biological interest. Moreover, other kinds of amino acids were also formed, nonbiological ones, a discovery that demonstrates the cosmic prebiological relevance of the experiments. This highly successful result ushered in a new era of experimental studies of prebiological chemistry, setting the stage for a decade of remarkable progress in the science.

Indeed, the 1950s were extraordinary not only for the strides made in prebiological chemistry but for the science of biochemistry in general, advancing our knowledge of the essence, dynamics, and chemical nature of living systems as well as our understanding of life's origin. In 1953, James Watson and Francis Crick proposed their double-helix model of DNA, the key to understanding heredity and the **genetic code.** Fred Sanger described the primary structure of a protein, Arthur Kornberg achieved in vitro synthesis of DNA, and Severo Ochoa and Marianne Grunberg-Manago discovered the specific enzyme—polynucleotide phosphorylase—required for the biosynthesis of RNA. The discovery in Ochoa's laboratory of the RNA-biosynthesizing enzyme polynucleotide phosphorylase was particularly important because it was followed by another, almost serendipitous, discovery. Marshall Niremberg and Heinrich Matthei, who had used polyU only as a control in their initial experiments, soon showed that the polynucleotide phosphorylase enzyme also catalyzes the biosynthesis of polyuridylic acid (polyU), the RNA messenger molecule responsible for building polyphenylalanine (linked monomers of the amino acid **phenylalanine**).

The discovery in Ochoa's laboratory and the breakthrough made by Niremberg and Matthei provided the key to unlock the genetic code.

Niremberg announced the way to crack the genetic code at the Fifth International Congress of Biochemistry, held in Moscow in 1961. At this same meeting, I also presented a paper, "Formation of Purines under Possible Primitive Conditions," an update of work carried out in 1959 (Oró 1960). This report offered a more complete story of our studies of the prebiotic synthesis of **adenine** and the **imidazole** intermediate compounds involved in its formation. My presentation built on a series of similar talks I had given in the United States the previous year.

PERSONAL REMINISCENCES

The problem of the origin of life has occupied my attention since my high school years, prior to and after the Spanish Civil War of 1936–39. The heart of the story is that I was searching for the meaning of human existence—and, I must say, my being the son of a baker helped me greatly in this endeavor.

As a youth in Lleida, Spain, I worked in my father's bakery. In our daily schedule, there was a free period of about an hour (from 4 to 5 o'clock in the morning) between the first and second batches of bread making. There was no point in trying to go back to sleep for such a short time. So, instead, I used the time to collect my thoughts, search my conscience, and quietly reflect. Often on clear, dark nights, I would gaze into the starry sky and ask myself the quintessential question: Is it possible that another baker's son is out there, on a planet circling one of the myriad stars of the universe, asking the same question I am asking now—"What in hell are we doing here?" Could I find the answer in books, or perhaps by asking one of my high school teachers? Fortunately, I had a good teacher in my biology class who sparked my interest in the life sciences and started me thinking about the complexity of the living process. My interest was further spurred by reading Spanish translations of Charles Darwin's *On the Origin of Species* (1859), Walther Löb's *Einführung in die Biochimie [Introduction to Biochemistry]* (1911), and Camille Flammarion's *La Pluralité des Mondes Habités [The Plurality of Inhabited Worlds]* (1917).

By the early 1940s, after the Spanish Civil War had ended and my father and I had rebuilt the family's ransacked business, it was time for me to decide the course of my life. One especially fine day, I was walking on the street near the bakery when it came, an answer to the existential question that had been bugging me. "I know!" I said to myself, "I'll devote my life to the study of the origin of life!" In retrospect, I suspect that my reading of the three books just noted led me in this direction. Darwin and Löb had taught me that both complex living systems and complex biomolecules evolved from simpler ones. And Flammarion had sparked my interest in comets and offered the tantalizing possibility that we may not be alone in the universe. As a direct consequence of this decision, I graduated in chemistry from the University of Barcelona in 1947, five years later. Some things changed: my parents passed away, I married. But my decision held firm.

In August 1952, on arriving in New York City for the first time, I visited Severo Ochoa in his laboratory at the New York University School of Medicine. Professor Ochoa invited me to lunch, and I used the opportunity to discuss with him my plans to study prebiotic chemical evolution. I asked him if it would be worthwhile to study nonenzymatic reactions involving formaldehyde and other simple organic compounds in the hope of discovering nonbiological pathways to the synthesis of the most important building blocks of biological macromolecules—amino acids, **purines,** and **pyrimidines.** His answer was wonderfully encouraging, providing support for the program of research I would eventually carry out after completing my doctoral studies in biochemistry. For my doctoral thesis at Baylor College of Medicine in Houston, I investigated the biochemical incorporation of ^{14}C-tagged **formate** (a salt of formic acid, $HCOOH$) into the purines that in part make up nucleic acids, and I studied the chemical mechanism of how formic acid is oxidized in living systems. During this period, I read Stanley Miller's now-famous 1953 publication. Though pleased to learn of this breakthrough, I was also shaken. I had not realized that anyone else was carrying out experiments on prebiotic chemistry and the origin of life. At the time, I knew nothing about the earlier experiments conducted by Melvin Calvin and his associates (Garrison et al. 1951; Calvin 1969).

A year or so before I graduated from Baylor, I learned from our dean that the department of chemistry at the University of Houston (UH) needed a biochemist and had requested my services. I journeyed across town, interviewed, and landed the job. As it turned out, they were in dire need—in my first year at UH, I was assigned to teach two lecture courses plus three laboratory courses. This left me little time to do research. But, inspired by Walther Löb's book and heartened by Severo Ochoa's encouragement, I did manage to start the experiments I had planned. My aim remained the same: to synthesize by nonbiological means relatively complex biochemical compounds from simple carbon- and nitrogen-containing starting materials.

Urea ($CO[NH_2]_2$) can be formed by heating **ammonium cyanate** ($NCONH_4$)—a reaction whose discovery by Friedrich Wöhler in the 1820s signaled the birth of organic chemistry (Leicester 1974). Moreover, in the presence of strong bases such as sodium hydroxide, formaldehyde (CH_2O) can condense with itself to generate a great diversity of sugars, a reaction first reported by Butlerow in the 1860s. Inspired by the Darwinian concept of "complexity from simplicity," I took my cue from these discoveries. I investigated two types of reactions. First, using two reactive compounds—formaldehyde (or the closely related **paraformalde-hyde**) and hydroxylamine-HCl—I showed that a number of amino acids and **hydroxy acids** could be made. Among the amino acids formed were **glycine, glycinamide, alanine, β-alanine, serine, aspartic acid,** and **threo-nine. Formic acid, acetic acid, glycolic acid,** and **lactic acid** were also formed. The mechanism of the reaction involved production of hydro-gen cyanide and ammonia as reacting intermediate compounds, and the first products formed were the corresponding amino acid amides, which yielded the amino acids and other acids on hydrolysis (Oró et al. 1959).

The second group of reactions I studied involved **hydrogen cyanide** (HCN), one of the major intermediates in Stanley Miller's amino acid–producing experiments, as well as those just described. Understanding this work depends on three pertinent facts:

1. **Cyanogen** (C_2N_2), the parent radical that gives rise to hydro-gen cyanide, is abundant both in comets and in the atmos-phere of the Sun, an abundance suggesting that hydrogen cyanide is probably widespread throughout the universe.

2. In 1875, the pioneering biochemist Eduard Pflüger had stated that "the beginning of organic life . . . is to be found to a very large degree in cyanogen. . . . [C]yanogen and its com-pounds had much time and opportunity to follow their great tendency for transformation and polymerization . . . into a labile protein which constitutes living matter" (see Lazcano 1995).

3. By 1948, my colleague Emilio Duró and I had lost all fear of handling industrial quantities of hydrogen cyanide, despite its

lethal toxicity. By that time, we had often used it as an intermediate reactant in the manufacture of commercial amounts of
various pharmaceuticals (for example, **mandelic acid**).

Thus, I was ready to attack a question raised at the First International
Congress on Oceanography held in New York at the United Nations
in 1959. The nature of the polymeric material formed in Miller's electric discharges experiments was as yet unstudied, but large amounts of
hydrogen cyanide had been produced. Was the polymer a product of
hydrogen cyanide?

Upon returning to Houston, I bubbled hydrogen cyanide gas into
an aqueous ammonia mixture. When I saw what was happening, it was
as though I had been transported back to primitive Earth. Pure alchemy
was taking place! Yellow bands or threads were formed as soon as the
gas entered the ammonium-charged solution. After a few minutes, the
color changed to orange, then reddish, then deep red, intensifying until it was jet black. Watching the changes, I was astonished, but I later
found that the final product, the deep black compound, was indeed
polymerized hydrogen cyanide.

The relevance of this result to the chemistry of the cosmos seems
clear. We now know that the black polymeric hydrogen cyanide synthesized in the laboratory is identical to the black dust that coats the
nuclei of comets, a blackness responsible for the low albedo of these solar system objects. This laboratory synthesis took place in a five-liter
round-bottomed flask; but in 1994 we were able to observe almost the
same process on a cosmic scale, when the comet **Shoemaker-Levy 9**
(SL-9) fell, in 22 pieces, into the upper atmosphere of Jupiter. First,
bright yellowish impact areas were observed, some as large as Earth;
then, after a few days, these areas turned dark brown-black. I was among
the fortunate observers permitted to see these brown-black areas in
Jupiter's atmosphere from the Pic-du-Midi observatory in France. I suggested to the astronomers there that the coloration in the impact areas
consisted primarily of polymeric hydrogen cyanide—an idea supported
by the subsequent findings of N. Gautier and other astronomers, who
have detected HCN and related compounds on the SL-9 spots on Jupiter.

Returning to experimental chemistry in the laboratory, we know that
hydrogen cyanide is highly reactive. It can exist in a chemically neutral
form or in an ionized form. Thus, it can combine with itself in several
different ways. And, depending on how it combines, it can generate many
different kinds of compounds—polymers of hydrogen cyanide, **amino**

acid nitriles, amides, adenine, and others. But the polymeric forms—and one in particular, adenine—held my attention. Indeed, that adenine, one of the two purines of DNA and RNA, is simply a polymer of hydrogen cyanide is well illustrated by its original name. One of the first published chemical analyses of this biologically important compound formally identified it as "pentameric hydrogen cyanide" (in German, *pentamere Cyansäure*) because its elemental composition, $H_5C_5N_5$, is exactly HCN multiplied by five.

In the HCN-polymerization experiment carried out in our laboratory in the summer of 1959, paper chromatography revealed the presence of amino acids, amino acid amides, and other acids that had been formed by hydrolytic breakdown of their corresponding nitriles. But by means of a selective Gerlach-Döring test, I also identified very small amounts of adenine. To confirm that adenine had indeed been formed in the reaction, I concentrated the products of the reaction and characterized adenine in several additional ways—first, by its ultraviolet (UV) absorption in a paper chromatogram, then by a specific Gerlach-Döring test. After isolating the compound from solution, I measured its UV absorption spectrum and determined the melting point of crystals prepared from one of its chemical derivatives, adenine picrate. All five of these tests confirmed that adenine had been synthesized.

The mechanism of adenine synthesis from hydrogen cyanide in aqueous solutions of ammonia is now well known (Oró 1960, 1961; Oró and Kimball 1961, 1962). And the synthesis of other purines (**guanine, xanthine,** and **hypoxanthine**) in the presence of simple compounds derived from hydrogen cyanide has also been studied in some detail. Moreover, adenine—pentameric hydrogen cyanide, the same jet black HCN polymer formed in the first set of experiments of 1959—is now known to be formed directly, even at low temperatures—a chemistry that explains why, as reported by Kissel and Krueger (1987), adenine and other purines have been detected by mass spectrometry in the dust from Halley's comet (see table 1.2). Adenine can also be produced from hydrogen cyanide and ammonia alone (Wakamatsu et al. 1966), by the UV polymerization of hydrogen cyanide, by irradiation with electrons (Ponnamperuma 1965; see also Oró 1965), and probably by the dehydrative polymerization of formamide, a mechanism first hypothesized by Leslie Orgel.

Results obtained from experimental studies of the synthesis of amino acids by formation of hydrogen cyanide from formaldehyde and other reactants (Oró et al. 1959), and of the synthesis of adenine and other

TABLE 1.2 ORGANIC MOLECULES DETECTED IN
COMET HALLEY DUST GRAINS

C, H Compounds	C, N, H Compounds	C, O, H Compounds
Pentyne	Hydrocyanic acid	Formaldehyde
Hexyne	Acetonitrile	Acetaldehyde
Butadiene	Propanenitrile	Formic acid
Pentadiene	Iminomethane	Acetic acid
Cyclopentene	Iminoethane	Isocyanic acid
Cyclohexene	Iminopropene	Methanol imine
Cyclohexadiene	Pyrrole, imidazole	Oximidazole
Benzene	Pyridine, pyrimidine	Oxypyrimidine
Toluene	Purine, adenine	Xanthine

Adapted from Kissel and Krueger (1987).

compounds from the polymerization of hydrogen cyanide (Oró 1960), led naturally to consideration of the role that HCN-rich comets may have played in forming organic biochemicals on primitive Earth (Oró 1961). According to the currently accepted hypothesis, the Earth–Moon system is the result of the impact of a Mars-sized body with the forming Earth (as discussed, for example, in chapter 2, and by Cameron and Benz 1991). In combination with the detailed analyses of the terrestrial planets summarized by Delsemme (2000), this hypothesis is consistent with the idea that most of the water in Earth's hydrosphere, together with the organic compounds that contributed to its earliest biosphere, came from cometary collisions with the primitive Earth. This perspective is not entirely new. In the early 1960s, I realized that the problem of the origin of life on Earth must be viewed not only Earth-centrically, but cosmochemically, as well. Indeed, our chemical studies at the University of Houston, sponsored by NASA for more than 30 years, were based on a broad cosmochemical perspective (Oró 1961, 1963). Table 1.3 summarizes the basis of our perspective— the chemical relations between cometary molecules and the biochemicals of which life is composed.

COSMOLOGICAL EVOLUTION AND THE ORIGIN OF LIFE

In recent years, discoveries of other planets, **protoplanetary disks,** and galaxies by the Hubble telescope have confirmed the idea that the cosmos is expanding and introduced a new dimension to our understanding of the origin of our Universe. This new knowledge is relevant to the problem of the origin of life.

TABLE 1.3 BIOCHEMICAL MONOMERS AND
PROPERTIES THAT CAN BE DERIVED FROM
COMETARY MOLECULES

Cometary Molecules	Formulas and Reactants	Biochemical Monomers and (Properties)
1. Hydrogen	H_2	(Reducing agent)
2. Water	H_2O	(Universal solvent)
3. Ammonia	NH_3	(Amination and catalysis)
4. Carbon monoxide	CO $(+H_2)$	Fatty acids
5a. Formaldehyde	CH_2O	Ribose and glycerol
5b. Aldehydes	$RCHO$ $(+HCN + NH_3)$	Amino acids
6. Hydrogen sulfide	H_2S (+other precursors)	Cysteine and methionine
7. Hydrogen cyanide	HCN	Purines and amino acids
8. Cyanoacetylene	HC_3N (+cyanate)	Pyrimidines
9. Phosphate[a]	PO_4^{3-}	Phosphates and nucleotides
10. Cyanamide[b]	H_2NCN	Oligonucleotides

[a]In interplanetary dust particles.
[b]Not yet detected in comets.

Major stages in the evolution of the cosmos are summarized in table 1.4, based on the currently favored **inflationary model** of cosmology (Guth 1994, 1997). Some 300,000 years after the **Big Bang,** the initial primordial explosion, free atomic nuclei and electrons condensed to form matter and the Universe became transparent to light energy (photons). At this point, cosmochemistry became pertinent to nuclear chemistry, to stellar and interstellar chemistry, and, over time, to the origin of life on Earth.

In the mid-1800s, Darwin showed how the concept of evolution by natural selection applies to living systems. But evolution also operates in the inanimate world, not only Earth but the Universe as a whole, including all cosmic bodies (galaxies, stars, circumstellar and interstellar clouds, interstellar molecules, planetary systems, planets, comets, asteroids, meteorites) and all chemical elements. Comets transported organic molecules and water to the primitive Earth early in the planet's history, presumably over a period of several hundred million years. In the oceans that then formed, both cometary and terrestrial (those synthesized directly in the early environment) organic molecules evolved by natural selection, ultimately giving rise to life—possibly in the "warm little pond" that Darwin envisioned in his famous letter to Joseph Hooker (see chapter 3). The linkage from cosmic elements to

TABLE 1.4 BIG BANG INFLATIONARY THEORY MODEL
OF THE UNIVERSE

	Time	Processes
Forces	10^{-43} seconds	Gravity separates from other forces
	10^{-33} seconds	Electroweak, strong nuclear, and gravity forces operate
	10^{-10} seconds	Weak nuclear and electromagnetic forces separate
Matter	10^{-6} seconds	Quarks combine to form particles
	3 minutes	Light atomic nuclei (such as H, He, Li) form
	300,000 years	Free nuclei and electrons condense to neutral atoms; Universe becomes transparent to photons
	>1 million years	H and He atoms form protogalaxies and stars

Adapted from Oró (1998); see also Guth (1994, 1997).

cometary molecules to primitive Earth to biological evolution ties cosmochemical evolution to the origin of life.

Nuclear Synthesis of the Biogenic Elements

From a chemist's point of view, one of the most creative processes in the evolution of the Universe was the nuclear synthesis of the elements. This synthesis took place in several stages—first, during the Big Bang as the Universe began, then in the cores of stars, and later in **supernova** explosions.

About three minutes after the Big Bang, free nuclei of hydrogen and helium were generated. After more than a million years had passed, giant clouds of hydrogen and helium condensed by gravitational attraction to form stars, which clustered into **protogalaxies.** In the hot cores of the newly formed stars, the various low-mass elements were synthesized. Most of the higher-mass elements were generated much later, typically in gigantic supernova explosions, which mark the end of a star's existence. As new stars are born and old stars die, the genesis of light and heavy elements continues.

Spectra gathered by astronomical observatories tell us that the Universe is composed almost entirely—about 98 percent—of hydrogen (H) and helium (He). The other elements (more than a hundred) amount to just 2 percent; but half of that small fraction, about 1 percent of the total, comprises carbon (C), oxygen (O), nitrogen (N), sulfur (S), and phosphorus (P). Thus, about 99 percent of the Universe is composed of helium plus the six biogenic elements, CHONSP. Thus, in terms of the origin of life, the genesis of the biogenic elements was probably the most important event in cosmochemistry. Table 1.5 summarizes the nuclear reactions that led to their nucleosynthesis (Macià et al. 1997).

Helium, Carbon, and Energy for Life

Helium nuclides, first synthesized during the Big Bang, are produced during star formation by a sequence of reactions known as the **proton–proton chain** (table 1.5). This sequence occurs in the cores of many ordinary stars, including the Sun, at about 15 million degrees Celsius. During this process, four protons condense together to form one nuclide of helium, a reaction that results in a small loss of mass. In accordance with Einstein's equation, $E = mc^2$, this mass is converted into a large amount of energy, which is released as heat and radiation.

The formation of carbon occurs in the interiors of stars at 100 million degrees Celsius. This reaction, producing one ^{12}C nuclide from the low-probability collision and condensation of three helium nuclides, also known as alpha (α) particles, is called the **triple-α process** (table 1.5). Once ^{12}C is made by the triple-α process, subsequent capture of α particles produces oxygen and sulfur nuclides. Nitrogen nuclides are generated catalytically in stars by a different process—the **CNO cycle**. But formation of nuclides of one biogenic element, phosphorus, requires

TABLE 1.5 NUCLEOSYNTHESIS OF THE
BIOGENIC ELEMENTS

Element	Source(s)	Production Processes	Cosmic Abundance (rank)
^{1}H	Big Bang	Primordial nucleosynthesis	1
^{2}He	Big Bang and hydrogen-burning	Primordial nucleosynthesis, proton–proton chain, and CNO cycle	2
^{6}C	Helium-burning	Triple-alpha process: $3\alpha \Rightarrow {}^{12}C$	4
^{7}N	Hydrogen-burning	CNO cycle	6
^{8}O	Helium-burning	Alpha capture: $^{12}C + \alpha \Rightarrow {}^{16}O$	3
^{10}Ne	Carbon-burning	Alpha capture: $^{16}O + \alpha \Rightarrow {}^{20}Ne$	5
^{16}S	Oxygen-burning	$^{16}O + {}^{16}O \Rightarrow {}^{28}Si + \alpha \Rightarrow {}^{32}S$	10
^{15}P	Carbon- and neon-burning	Many processes	17

Adapted from Macià et al. (1997).

many complex nuclear reactions (Macià et al. 1997). This more diffi-cult synthesis explains why phosphorus is less abundant in the cosmos than the other principal elements of life.

Circumstellar and Interstellar Molecules

Carbon stars are rich sources of carbon compounds and organic mole-cules. Formed in the hot cores of these and other stars, the biogenic elements migrate from the interiors to the cooler outer regions. There, more ordinary chemical reactions give rise to diatomic and triatomic elemental combinations, producing compounds that can be observed in stellar atmospheres. Among the most common molecular species detected are C_2, CN, CO, CH, NH, OH, and H_2O, all of which are present in the atmospheres of ordinary **main-sequence stars** such as our Sun.

More than one hundred chemical species have been identified in the interstellar medium by their distinctive gas-phase molecular spectra (Oró and Cosmovici 1997; Oró 2000). All these molecules, ions, and radicals are relatively simple, for the most part comprising just a few (two to four) atoms; among the largest detected is the 13-atom com-pound $HC_{11}N$. About 75 percent of the various species now known are organic, in that they contain atoms of carbon or carbon linked to hydrogen. The biogenic elements are the most abundant reactive ele-ments in the Universe (helium, although highly abundant, is chemically inert); moreover, most known interstellar molecules contain carbon. Thus, we can fairly say that the essence of the Universe is organic, fully prepared for life to emerge wherever and whenever conditions permit.

Of the currently identified interstellar species, ten compounds are espe-cially relevant to life: diatomic hydrogen (H_2), ammonia (NH_3), water (H_2O), carbon monoxide (CO), formaldehyde (CH_2O), hydrogen sulfide (H_2S), hydrogen cyanide (HCN), cyanoacetylene (HC_3N), phosphorus nitrile ($P[CN]_5$), and cyanamide (H_2NCN). Using these compounds as starting materials, one could synthesize in the laboratory, under plausi-ble prebiological conditions, amino acids, nucleic acid bases, sugars, lipids, and **mononucleotides** of the kinds present in all living systems (table 1.3). But without such starting materials, there could be no life.

THE EARTH-MOON SYSTEM

About 5,000 million years ago, the Solar System formed by the gravi-tational collapse of a dusty gaseous nebula of interstellar matter. Pre-sumably, this collapse was triggered by the shock wave of a nearby supernova explosion, as suggested by studies of Hoppe and colleagues (1997), who have shown that grains of **silicon carbide** (SiC) in the Murchison meteorite have isotopic compositions indicating supernova origin. The first several million years of Solar System evolution involved great upheaval, with the bodies of the system more or less continuously colliding with **planetesimals,** comets, and other major objects as they orbited the forming Sun. Thus, the early history of the Solar System can be divided into two phases—first, the gravitational collapse of the solar nebula to form the **proto-Sun;** second, the accretion of differentiated nebular matter into planets, satellites, and other bodies. As Delsemme (2000) has pointed out, the precursors of the innermost, **terrestrial planets** (Mercury, Venus, Earth, and Mars) were rocky planetesimals essentially devoid of low-mass molecules, owing to the relatively high temperatures prevailing in the inner part of the **asteroidal belt** (between Mars and Jupiter) whence they derived.

The record of impacts preserved on the surface of the Moon bears witness to the turbulent state that prevailed during the early stages of formation of the Solar System. **Proto-Earth,** too, must have undergone many collisions with planetesimals, small and large. Cameron and his colleagues have proposed a new (and, in some ways, revolutionary) model for the formation of the Earth–Moon system that explains how the system may have come into being (see, for example, Cameron and Benz 1991). According to this theory, a celestial body having a mass comparable to that of Mars collided with the proto-Earth, injecting into the Earth much of the iron now present in its core. As a result of this fusion, portions of Earth's rocky mantle were ejected into orbit and coalesced to form the Moon, the largest satellite encircling any of the terrestrial planets. This theory explains most of the similarities between the composition of Earth's mantle and the Moon, as well as the slight differences—for example, the Moon's lower iron content. It also explains various dynamic aspects of the system, such as the angular momentum of the Earth–Moon pair.

Cometary Matter Captured by the Early Earth

During the catastrophic collision posited by Cameron's theory, most of the gaseous volatiles the colliding bodies originally harbored would have been lost to space. If so, the inventory of such volatiles—water and organic compounds on which the origin of life would ultimately depend—must have been replenished, contributed by comets and other small bodies of the evolving Solar System. These bodies would have continued to bombard Earth during the late stages of planetary accretion, the first 600 million years of the planet's history (Cameron and Benz 1991). Earth would have retained such volatiles gravitationally, allowing its atmosphere and hydrosphere to form; but no such retention would have occurred on the Moon because of its much lower mass. (Today the Moon's total atmospheric pressure is 10^{-10} torr, less than a trillionth of that on Earth.) While some collisions would have contributed water and volatile compounds to Earth, a process known as impact capture, others would have removed such materials, sending them into space (impact erosion). But by the end of this period of early bombardment, as extrapolated from the cratering record on the Moon, Earth's mass would have significantly increased (table 1.6)—most notably for the origin of life, in its content of water, as well as carbon and the other biogenic elements. (Such a process may have occurred repeatedly throughout the Solar System, not only on Earth, but also on the other major terrestrial planets, Jupiter's satellite **Titan,** and elsewhere.)

Comets are very low-temperature aggregates of interstellar matter that are gravitationally attracted by the Sun. Originating both in interstellar space (in the **Oort cloud**) and in outer regions of the Solar System (the **Kuiper belt**), they are regarded as huge "dirty snowballs" (a model that derives from the pioneering studies of the astronomer Fred Whipple). Delsemme has studied the composition of comets and

TABLE 1.6 COMETARY MATTER TRAPPED BY SOLAR
SYSTEM BODIES

Body	Cometary Matter (g)	Time Span
Venus	4.0×10^{20}	2.0×10^9 years
Moon	2.0×10^{20}	During late accretion
Earth	2.0×10^{14} to 2.0×10^{18}	2.0×10^9 years
	1.0×10^{25} to 1.0×10^{26}	During late accretion
	3.5×10^{21}	During late accretion
	7.0×10^{23}	4.5×10^9 years
	2.0×10^{22}	4.5×10^9 years
	1.0×10^{23}	2.0×10^9 years
	1.0×10^{24} to 1.0×10^{25}	1.0×10^9 years
	6.0×10^{24} to 6.0×10^{25}	1.0×10^9 years
	1.0×10^{23} to 1.0×10^{26}	4.5×10^9 years

Adapted from Oró et al. (1992).

their relations to the terrestrial planets in detail (see Delsemme 1992), while Greenberg and Hage have developed a model that explains their origin from the condensation of interstellar grains to the formation of ever-larger aggregates (Greenberg and Hage 1990).

In recent years, a number of comets have been studied in considerable detail—Halley's (Kissel and Krueger 1987), Shoemaker-Levy 9 (by N. Gautier at the Pic-du-Midi observatory), Hyakutake (Mumma 1997), and, of course, the spectacular Hale-Bopp of 1997. So far, more than 33 chemical species have been identified in Hale-Bopp, most of which are the same organic and inorganic molecules detected in the interstellar medium. The organic composition of comets is probably best reviewed by Mumma (1997), who was the first to use high-resolution infrared spectroscopy to analyze the Hyakutake comet. He and his colleagues detected strong emissions from H_2O, HDO, CO, CH_4, C_2H_2, C_2H_2, CH_2OH, H_2CO, OCS, HCN, OH, and other chemical species. Of particular relevance to the origin of life are the large amounts of methane (CH_4) and ethane (C_2H_6) detected in the nucleus of Hyakutake. The relation of comets to the emergence of life on Earth has been discussed recently by Oró and Lazcano (1997), Oró and Cosmovici (1997), and Oró (2000).

Comets offer us a bridge that connects the chemistry of life on Earth to that of the Universe at large. This bridging function may be one of the most important consequences of cosmochemical evolution. I proposed comets as a source of the organic molecules necessary for the development of life on Earth many years ago in my paper "Comets and the Formation of Biochemical Compounds on the Primitive Earth" (Oró 1961). A major study that confirms the relevance of this notion was published by Armand Delsemme (2000): "The Cometary Origin of the Biosphere." (For additional discussion of the role of cometary organic matter in the origin of life, see Oró 2000.)

Prebiotic Chemical Evolution

I have discussed a number of physical and chemical creative processes in the history of the Universe that are relevant to the origin of life. These creative processes have continued from the Big Bang through the formation of stars and galaxies, the nuclear syntheses of biogenic elements in stars and supernova explosions, the formation of molecules in circumstellar and interstellar clouds, and, eventually, the formation of planetary systems like our own.

TABLE 1.7 PREBIOTIC SYNTHESIS OF BIOCHEMICAL COMPOUNDS FROM COMETARY MOLECULES

Precursor Molecules	Reactions	Monomers	Polymers	Potential Product
HCN	→ (1) →	amino acids →	peptides →	PROTOCELL
HCN	→ (2) →	purines →	nucleotides →	PROTOCELL
HC_3N	→ (3) →	pyrimidines →	nucleotides →	PROTOCELL
CH_2O	→ (4) →	ribose (5-p) →	nucleotides →	PROTOCELL
$CO + H_2$	→ (5) →	fatty acids (glycerolphosphate + choline) →	phospholipids (liposomes) →	PROTOCELL

(1) $+ H_2O + NH_3$
(2) $+ H_2O + NH_3$
(3) $+ H_2O$ + urea
(4) + P (phosphate)
(5) Fischer–Tropsch catalysis
Adapted from Oró (2000).

Once Earth acquired water and the cometary precursors needed for nonbiological synthesis of biochemicals, life could not have emerged without conditions that allowed synthesis of biochemical **monomers** and **polymers** (**polynucleotides, polypeptides,** and **lipids**)—biochemicals whose self-organization and natural selection could give rise to self-reproducing living systems capable of Darwinian evolution. Such processes of prebiological chemical evolution are addressed in the following chapters, and have been described in a number of papers from different laboratories, including our own. Recent reviews can be found in Oró et al. (1990) and Oró (2000). An overall summary of the problem, based largely on the importance of comets for the synthesis of the essential biochemical molecules to life, is shown in table 1.7.

Biological Evolution and Mass Extinctions

Life has evolved on Earth over an unimaginably long time—4,000 million years or more. The oldest fossils now known are those found in rocks from Western Australia (the **Apex chert**), estimated to be nearly 3,500 million years old. Many of the microscopic fossils found, 11 different species, have morphologies similar to those of photosynthetic oxygen-producing **cyanobacteria** that live today (Schopf 1993). The evolutionary process from these ancient microbes to present-day species, studied so patiently by Darwin, is a near-miracle that is beyond the scope of this chapter. Suffice it to say, quoting Darwin's *On the Origin of Species* (1859): "From so simple a beginning, endless forms most beautiful and most wonderful have been, and are being evolved." And I say that, among this endless series of "most beautiful and most wonderful" forms, are microbes, flowering plants, dinosaurs, and hominids, including ourselves, *Homo sapiens sapiens*. (Additional discussion of life's origin and earliest evolution can be found in two excellent books: Miller and Orgel 1974; Schopf 1999.)

But, as Carl Sagan liked to remind us, "Comets giveth, and comets taketh away." One important inflection point in the Darwinian evolutionary process took place about 65 million years ago with the impact of a comet (or possibly an asteroid) in Mexico's Yucatán Peninsula. This comet, which left its mark as the Chicxulub crater, one of the largest on the planet, was first suggested by discovery of abnormal amounts of **iridium,** a cosmic marker in geologic units of this age. This impact caused a catastrophic darkening of the Earth, presumably resulting in the extinction of many kinds of life—including dinosaurs, the dominant biologic group for nearly 200 million years. These extinctions opened many new ecologic niches, some of which were exploited by warm-blooded animals like ourselves, mammals. The rise of mammals eventually led to the appearance of primates—and, through them, to the rise of humans—several million years ago.

A similar cosmic event, although without such Earth-shattering consequences, happened not long ago—the collision of comet Shoemaker-Levy 9 with Jupiter. Had any such collision occurred with the Earth, it would almost certainly have wiped out much of the biosphere, including humans (although various microorganisms might be expected to survive, especially those living in the seas at abyssal depths). Indeed, the "mother of all extinctions" (Luann Becker, quoted in Golden 2001), which wiped out more than 90 percent of oceanic invertebrate species and most genera of land plants and vertebrates about 250 million years ago, is attributable to such a collision. This extinction was probably triggered by the impact of an asteroid or comet some 6 to 12 km across (Becker et al. 2001). That cosmic body would have slammed into Earth with the force of a magnitude-12 earthquake (Golden 2001). According to some investigators (Ward and Brownlee 2000), events such as these may be of serious concern for the future of intelligent life on our planet.

EXPLORATION OF THE SOLAR SYSTEM

The Apollo Program

On July 20, 1969, Neil Armstrong became the first human to set foot on the Moon—the only heavenly body yet visited by humans. The lunar samples he and Buzz Aldrin collected, the moon dust, were notably blackish. At the time, the Houston newspapers reported that the material brought back could be graphite, a coal-like crystalline carbon. This preliminary impression kindled hopes that the Moon might have harbored life in the distant past. These hopes died after analysis of the lunar matter. Today, we know that life could never have existed on the Moon, and that the dark color of the lunar dust was mainly due to constant irradiation by the **solar wind.** In our laboratory, GC-MS (**gas chromatography–mass spectrometry**) analyses confirmed that, with the exception of a tiny amount (a few parts per million) of carbon monoxide, methane, and a few other simple compounds, the lunar samples were devoid of organic matter. Others reported finding traces of amino acids, but these were presumably products of reactions of hydrogen cyanide brought to the Moon by comets, or of carbon, nitrogen, and hydrogen atoms implanted on the Moon by the solar wind.

Life on Mars and Europa?

Even before the first Apollo flight, NASA had targeted Mars as the next goal for space exploration. So, on July 20, 1976, exactly seven years after the initial Apollo flight, the first of two NASA Viking spacecraft landed on the surface of Mars. True to its sobriquet, the Red Planet was soon seen to be covered by reddish-brown dust. As the first pictures were received at the Jet Propulsion Laboratory in Pasadena, California, I was on hand as a member of the Viking molecular analysis team. There, Dwayne Anderson, a colleague on the team, reminded me of my own prediction—that the redness suggested a highly oxidized soil, inhospitable to life.

Soon thereafter, our team began the hunt for volatizable organic and inorganic compounds in the Martian dust. For molecular analysis, we used Viking's GC-MS instrument, which was similar, in principle, to the one used earlier in our laboratory for studies of the lunar samples. Viking's new miniaturized apparatus was built according to suggestions that Klaus Biemann and I had offered, together with other members of our team. It differed from our laboratory instrument in one significant respect—its weight, just 20 kilograms instead of 2 tons. What were our findings? We found that no organic matter was present, not even at the level of parts per billion, at either of the landing sites, the plains of Chryse and Utopia (Biemann et al. 1976). No evidence of life.

In contrast with these findings, one of the three sets of biological experiments conducted by the Viking spacecraft showed that incubation of the Martian dust in a nutrient solution containing ^{14}C-tagged formic acid ($H^{14}COOH$) resulted in rapid generation of radioactive carbon dioxide ($^{14}CO_2$). Levin and Straat (1976) interpreted this result as evidence of active microbial metabolism. I countered that the carbon dioxide did not come from microbes, but was the product of a chemical reaction in which the mineralic iron oxides (the source of the dust's red color) and hydrogen peroxide (H_2O_2) present in the analyzed samples had oxidized $H^{14}COOH$ to $^{14}CO_2$. This was 1976, but my assessment was based on research I had done some 20 years earlier while working on my doctoral thesis. I had studied the biological mechanism of formic acid oxidation in living systems, and I knew, even at the time, that in the presence of hydrogen peroxide, formic acid is rapidly oxidized to carbon dioxide. This oxidation occurs biologically, catalyzed by enzymes such as catalase (Oró and Rappoport 1959), which have an iron atom at their reactive center; but it is also catalyzed by inorganic iron (Fe^{2+}, Fe^{3+}), which is abundantly present in the iron oxide minerals hematite (Fe_2O_3) and magnetite (Fe_3O_4) on the Martian surface. Moreover, laboratory studies have since shown that the high flux of ultraviolet light impinging on the surface of the Red Planet would have rapidly oxidized, and thus destroyed, any organic matter that might have been present initially. For example, the half-life of residence of any meteoritic or cometary organic matter exposed on the surface of Mars is a scant few months (Oró and Holzer 1979), a mere instant in geologic time.

Although we found no evidence of life at the two Viking landing sites, we cannot rule out the possibility that life might once have existed on Mars in the distant past. Thus, it remains worthwhile to revisit the planet to look for such life in the form of buried and preserved fossils and their organic remnants. But what of Martian meteorite ALH 84001 and the possible presence of fossilized relics of microbial life some 3,600 million years old (McKay et al. 1996)? I can add nothing to what we have said before (Oró and Cosmovici 1997). Whether life did or does exist on Mars, we will have to wait for certainty until samples are brought to Earth and studied in the laboratory.

In addition to Mars, other targets in the Solar System, notably Jupiter's satellite **Europa,** have raised interest, both in the scientific community and within NASA, as possible abodes of present or past microbial life. NASA's Galileo spacecraft has obtained beautiful images of an array of linear canyons that crisscross the surface of Europa. These images confirm the frozen and cracked nature of Europa's icy surface, which is thought to overlie a global ocean—a prime site, as we suggested some years ago, for the search for life (Oró et al. 1992b). But the surface itself is evidently inhospitable, as suggested by spectrophotometric findings of hydrogen peroxide and concentrations of sulfuric acid.

LIFE BEYOND THE SOLAR SYSTEM

Although it sounds like science fiction, it is perfectly reasonable to imagine that life exists in other planetary systems. Particularly promising is the orbital system that encircles the star Beta-Pictoris, some 54 light-years distant from Earth. Beta-Pictoris is intriguing because it is surrounded by a vast disk of comet-like matter—as we have seen, a rich source of life-generating elements and organic compounds. Owing to the intense infrared radiation it emits, this disk was first detected by NASA's infrared astronomical satellite (IRAS), then photographed by American astronomers at the Las Campanas Observatory in Chile. Studies of Beta-Pictoris conducted by French astronomers suggest that at least 100 comets fall onto the central star every year. But this estimate is a minimum. Indeed, more recent work has shown that an infall of about 1,000 comets per year, each about 1 km across, would be necessary to explain the observed disappearance of cometary dust from the central area of the star-encircling disk where planets may be present (Lagage and Pantin 1994). If the orbital system around Beta-Pictoris includes Earth-like planets, life may have already emerged, or be emerging now, on one or more of them. And Beta-Pictoris is not the only promising site for the search for life: There is considerable indirect evidence that other nearby stars are orbited by planetary companions, as well as direct evidence of planet-forming, protoplanetary disks in the Orion Nebula. Such protoplanetary disks have been detected around approximately half of the stars studied in that nebula.

Discovery of Extrasolar Planets

Given the existence of other Earth-like planets, the emergence of life, even of intelligent life, is possible. But it may or may not be common throughout the Universe (Drake 1963; Oró 1995; Ward and Brownlee 2000). Moreover, we do not know whether Earth-like planets are common, as some researchers assert, or vanishingly rare, as others have claimed. The detection of small, Earth-like bodies is difficult, especially if they are far distant from the Earth, but the search for extrasolar planets, actively encouraged by NASA, has already yielded promising results.

In one of the major scientific discoveries of the twentieth century, numerous planets—all of them 50 or so light-years distant from our Solar System—have been detected within the past decade. The first, discovered by Michael Mayor and Didier Queloz of the Geneva Observatory in Switzerland, was a huge planet orbiting star 51 Pegasi B (Mayor et al. 1997). Since this breakthrough, nearly 50 other planetary systems have been detected (many reviewed by Marcy and Butler 1999). Among the more interesting are a multiplanet system encircling the star Upsilon Andromedae (Butler et al. 1999) and two somewhat smaller planets, a bit less massive than Saturn, orbiting stars HD 46375 and 79 Ceti. Because of the detection system used, only exceptionally massive planets, those about as large as or larger than Jupiter or Saturn, have been discovered so far, but their very existence increases hopes that smaller, Earth-like bodies will ultimately be found.

For the time being, however, discovery of extrasolar planets having orbits and masses comparable to those of Earth will have to await the deployment of the large infrared interferometry telescopic array currently under development by NASA. The first indicators of the existence of life on such extrasolar planets may come from studies of atmospheric chemistry, in particular by spectral detection of appreciable quantities of diatomic, molecular oxygen (O_2, not produced by volcanic or other geologic processes), **ozone** (O_3, a triatomic molecule produced photochemically from O_2), or perhaps water vapor (H_2O, without which life as we know it could not exist). If such data are forthcoming, they will provide a much improved basis on which to calculate statistically the possibility, or perhaps probability, of the existence of extraterrestrial intelligence. It is, of course, difficult to guess about such matters, but it seems likely that we will have to wait several decades before more telling information bearing on these questions becomes available.

The Search for Extraterrestrial Intelligence (SETI)

At present, the existence of extraterrestrial life, not to mention extraterrestrial intelligence, is an open question. It is even more difficult to say whether technologically advanced civilizations, capable of transmitting and receiving intelligible signals, exist in other planetary systems. Some years ago, Frank Drake (1963) developed an equation composed of a series of variables relevant to the problem. Using this equation, he calculated that our own **Milky Way Galaxy** may boast one advanced planetary civilization for every 10 million or so stars. However, the galaxy contains, literally, an astronomical number of stars; thus, his calculation means that civilizations may be rare and sporadically distributed but, at the same time, numerous.

How can we resolve this enigma? Several researchers, Drake among them, have suggested that we cannot use manned spaceships to explore even our own galaxy. The voyage of a single spacecraft to a single nearby star would consume an amount of energy equivalent to that produced by all humans on Earth. But some years ago, an interested group of American researchers—most prominently, Frank Drake, Don Morrison, and Jill Tarter, as well as two who have since passed away, Barney Oliver and Carl Sagan—suggested another way to attack the problem. They proposed that instead of actively seeking such civilizations by journeying to distant stars, it would be wiser to tackle the problem passively, by simply listening on appropriate channels for intelligent radio signals emitted by other civilizations.

Consequently, a radio signal–based search for extraterrestrial intelligence (**SETI**) began in 1992; it was supported by NASA at a fraction of the cost of any other space project. After just a few years, however, federal funding was withdrawn, largely owing to the objections of Senator William Proxmire (who, it is said, claimed to be unsure that there were intelligent beings in Washington, D.C., much less in outer space). In spite of this setback, the search has continued as a venture playfully renamed Project Phoenix, under the auspices of Frank Drake's privately funded SETI Institute in northern California. An active complement to this effort has been launched by Guillermo Lemarchand (2000) and his research team in Argentina. Although no extraterrestrial intelligent signals have been identified, the search goes on. But astronomical distances are large, and the number of technological civilizations in the Universe may be small, so a long time may pass before any signals are received (Oró 1999).

EPILOGUE

The Universe contains mostly hydrogen and helium, but it is also rich in organic compounds that are precursors of key biochemical molecules conducive to the emergence of life. And, surprisingly, some simple biochemicals involved in the transmission of nerve impulses (**neurotransmitters**)—the amino acids **glycine, glutamic acid,** and **γ-aminobutyric acid,** for example—have been found in the Murchison meteorite, thus demonstrating their extraterrestrial origin. One could therefore say that the Universe is ready not only for the emergence of life but also for the appearance of intelligence.

Conceivably, then, the Universe may harbor civilizations more intelligent than our own. Perhaps one day, through interstellar communication, some advanced civilization will help us resolve such age-old problems as war, famine, disease, overpopulation, misuse of natural resources, and human aging. But until utopia becomes reality, we would do well to cherish our own small blue planet, with all its varied and wonderful forms of life. We are all children of the Universe, but our Solar System owns only one Earth.

Peace from Cosmic Evolution

The Apollo astronauts' landing on the Moon allowed them to see the Earth as a small distant body enveloped by the immensity of space. They did not see the borders that separate people into different nations, nor the other things that separate us, such as the color of people's skin. Far beyond the Earth, the astronauts could see the world as one, an almost insignificant blue globe populated by brothers and sisters of the human family, who, at their wisest and best, have the ability to share their planet's limited resources and might yet learn to live in peace. Indeed, the astronauts' message, sent to us here on Earth, was simply, "We came in peace representing all humankind." This plain, yet profound reflection can be extended to embrace a number of ethical principles derived from a better understanding of the cosmos and the origin of life on Earth (Oró 1998):

1. Humility: The life of all cells descends from simple molecules.

2. Solidarity: Our genes have a common origin *(Homo sapiens)*.

3. Cooperation: We live on a resource-limited planet.

4. Hope: Someday we may communicate with more advanced civilizations.

5. Universality: We come from stardust and to stardust we shall return.

6. Peace: We should change our culture of war into a culture of peace.

7. Golden Rule: We should treat others as we would like them to treat us.

ACKNOWLEDGMENTS

I am pleased to salute Meritxell Cos Badia, whose interest and hard work were invaluable in the preparation of this article. Deepest gratitude also goes to the other participants in the UCLA Gold Medal Symposium on the Origin of Life—particularly J. William (Bill) Schopf and Jane Shen Schopf, who, with the help of Richard Mantonya, organized this memorable gathering. There is no question that the yearly all-campus symposia convened by CSEOL (UCLA's Center for the Study of Evolution and the Origin of Life) render a great service not only to UCLA, but to the larger community as well. Only a well-informed society will be able to face and resolve the challenges the twenty-first century will bring. Let us hope that, with humility, solidarity, and cooperation, we will be able to manage and wisely distribute the limited resources of planet Earth, protecting our terrestrial environment, so that future generations may live harmoniously and peacefully.

REFERENCES

Bernal, J. D. 1939. *The social function of science.* London: Routledge and Sons.
——. 1949. The physical basis of life. *Proceedings of the Royal Society of London* 357A:537–58.
——. 1951. *The physical basis of life.* London: Routledge and Paul.
——. 1958. *World without war.* New York: Monthly Review Press.
——. 1967. *The origin of life.* Cleveland, Ohio: World Publishing.
Biemann, K., J. Oró, P. Toulmin III, L. E. Orgel, A. O. Nier, D. M. Anderson, P. G. Simmonds, D. Flory, A. V. Diaz, D. R. Rushneck, and J. A. Biller. 1976. Search for organic and volatile inorganic compounds in two surface samples from the Chryse Planitia region of Mars. *Science* 194:72–76.
Butler, R. P., G. W. Marcy, D. A. Fischer, T. W. Brown, A. R. Contos, S. G. Korzennik, P. Nisenson, and R. W. Noyes. 1999. Evidence for multiple companions to Upsilon Andromedae. *Astronomical Journal* 526:916–27.
Calvin, M. 1969. *Chemical evolution.* Oxford: Oxford University Press.
Cameron, A. G. W., and W. Benz. 1991. The origin of the Moon and the single impact hypothesis IV. *Icarus* 92:204–16.
Cronin, J. R., and S. Pizzarello. 1997. Enantiomeric excesses in meteoritic amino acids. *Science* 275:951–55.

Darwin, C. 1859. *On the origin of species by means of natural selection*. London: John Murray.

Delsemme, A. 1992. Cometary origin of carbon, nitrogen and water on the Earth. *Origins of life* 21:279–98.

Delsemme, A. 2000. The cometary origin of the biosphere. *Icarus* 146:313–25.

Drake, F. D. 1963. The radio search for intelligent extraterrestrial life. In *Current aspects of exobiology*, ed. G. Mamikunian and M. H. Briggs, 323–45. Pergamon: New York.

Eccles, J. C. 1991. *Evolution of the brain: Creation of the self*. London: Routledge.

Flammarion, C. 1917. *La pluralité des mondes habités [The plurality of inhabited worlds]*. Paris: E. Flammarion.

Garrison, W. H., D. C. Morrison, J. G. Hamilton, A. A. Benson, and M. Calvin. 1951. The reduction of carbon dioxide in aqueous solutions by ionizing radiation. *Science* 114:461.

Golden, R. 2001. Great balls of fire. *Time*, 5 March, 59.

Greenberg, J. M., and J. I. Hage. 1990. From interstellar dust to comets: A unification of observational constraints. *Applied Journal of Science* 361: 260–74.

Grobstein, C. 1965. *The strategy of life*. San Francisco: Freeman.

Guth, A. H. 1994. The Big Bang and cosmic inflation. In *The Oskar Klein Memorial Lectures* 2, ed. G. Ekspong, 27–70. Stockholm: World Scientific.

Guth, A. H. 1997. *The Inflationary Universe*. Reading, Mass.: Addison-Wesley.

Hoppe, P. et al. 1997. Type II supernova matter in a silicon carbide grain from the Murchison meteorite. *Science* 272:1314–17.

Kissel, J., and F. R. Krueger. 1987. The organic component in dust from comet Halley as measured by the PUMA mass spectrometer on board Vega 1. *Nature* 326:755–60.

Lagage, O. O., and E. Pantin. 1994. Dust depletion in the inner disk around Beta-Pictoris as a possible indicator of planets. *Nature* 369:628–30.

Lazcano, A. 1992. *The spark of life, Alexander I. Oparin*. Mexico City: Pangea Editores.

Lazcano, A. 1995. A. I. Oparin: The man and his theory. In *Evolutionary biochemistry and related areas of physicochemical biology*, ed. B. F. Poglazov, B. I. Kurganov, M. S. Kritsky, and K. L. Gladilin, 49–56. Moscow: A. N. Bach Institute of Biochemistry and ANKO Press.

Leicester, H. M. 1974. *Development of biochemical concepts from ancient to modern times*. Cambridge: Harvard University Press.

Lemarchand, G. 2000. A new era in the search for life in the Universe. In *Bioastronomy '99: A new era in bioastronomy*, ed. A. G. Lemarchand and K. J. Meech, 7–18. San Francisco: Astronomical Society of the Pacific.

Levin, G. V., and P. A. Straat. 1976. Viking labeled release biology experiment: Interim results. *Science* 194:1322–29.

Löb, W. 1911. *Einführung in die biochimie [Introduction to biochemistry]*. Leipzig: B. G. Teubner.

Macià, E., V. Hernandez, and J. Oró. 1997. Primary sources of phosphorus and phosphates in chemical evolution. *Origins of life and evolution of the biosphere* 27:459–80.

Marcy, G. W., and R. P. Butler. 1999. Extrasolar planets: Techniques, results, and the future. In *The origin of stars and planetary systems,* ed. C. J. Lada and N. D. Kylafis, 681–708. Dordrecht, The Netherlands: Kluwer.

Margulis, L., and D. Sagan. 1997. *What is life?* Barcelona, Spain: Proa (in Spanish).

Mayor, M., D. Queloz, S. Udry, and J. L. Halbachs. 1997. From brown dwarfs to planets. In *Astronomical and biochemical origins and the search for life in the Universe,* ed. C. B. Cosmovici, S. Bowyer, and D. Werthimer, 313–30. Bologna: Editrice Compositori.

McKay, D. S., E. K. Gibson Jr., K. L. Thomas-Keprta, H. Vali, C. S. Ramaneck, S. J. Clemett, X. D. F. Chiller, C. R. Maechling, and R. N. Zare. 1996. Search for past life on Mars: Possible relic biogenic activity in Martian meteorite ALH84001. *Science* 273:924–30.

Miller, S. L. 1953. A production of amino acids under possible primitive Earth conditions. *Science* 117:528.

Miller, S. L. 1957. The mechanism of synthesis of amino acids by electric discharges. *Biochimica et Biophysica Acta* 23:480.

Miller, S. L., and L. E. Orgel. 1974. *The origins of life on the Earth.* Englewood Cliffs, N. J.: Prentice-Hall.

Moreno, A. 1998. Definition of life. Questions and problems. In *Life in the Universe. Horizons and frontiers, Summer Studies Course G.3C,* 92–106 (in Spanish). San Sebastian, Spain: Basque Country University Publications.

Mumma, M. J. 1997. Organics in comets. In *Astronomical and biochemical origins and the search for life in the Universe,* ed. C. B. Cosmovici, S. Bowyer, and D. Werthimer, 121–42. Bologna: Editrice Compositori.

Oparin, A. I. 1924. *Proiskhozhedenie Zhizni* (Moskva: Moskovskii Rabochii). [Reprinted and translated into English by J. D. Bernal, 1967. *The origin of life,* 199–234. London: Weidenfeld and Nicolson.]

———. 1938. *The origin of life.* New York: Macmillan.

———. 1986. Life: One of the forms of movement of matter. In *Origin of life and evolution of cells. Tributes of the society of Catalonian biology,* ed. L. Margulis, R. Guerrero, and A. Lazcano, 39:15–36 (in Catalan).

Orgel, L. E. 1973. *The origins of life.* New York: Wiley.

Oró, J. 1960. Synthesis of adenine from ammonium cyanide. *Biochemical and Biophysical Research Communications* 2:407–12.

———. 1961. Comets and the formation of biochemical compounds on the primitive Earth. *Nature* 190:389–90.

———. 1963. Studies in experimental organic cosmos chemistry. *Annals of the New York Academy of Sciences* 108:464–81.

———. 1965. Stages and mechanisms of prebiological synthesis. In *The origin of prebiological systems,* ed. S. W. Fox, 137–71. New York: Academic Press.

———. 1995. The chemical and biological basis of intelligent terrestrial life from an evolutionary perspective. In *Progress in the search for extraterrestrial life, ASP Conference Series 74,* ed. G. Seth Shostak, 121–33.

———. 1998. Cosmochemical evolution. A unifying and creative process in the Universe. In *Exobiology: Matter, energy, and information in the origin and evolution of life in the Universe,* ed. J. Chela-Flores and F. Raulin, 11–32. Amsterdam: Kluwer.

———. 1999. Evolutionary requirements for development of intelligent life. In *Origins of intelligent life in the Universe,* ed. R. Colombo, G. Giorello, G. Rigamonti, E. Sindoni, and C. Sinigaglia, 167–95. Como, Italy: Edicioni.

———. 2000. Organic matter and the origin of life in the Solar System. In *Bioastronomy '99: A new era in bioastronomy,* ed. A. G Lemarchand and K. J. Meech, 285–302. San Francisco: Astronomical Society of the Pacific.

Oró, J., and C. B. Cosmovici. 1997. Comets and life on the primitive Earth. In *Astronomical and biochemical origins and the search for life in the Universe,* ed. C. B. Cosmovici, S. Bowyer, and D. Werthimer, 97–120. Bologna: Editrice Compositori.

Oró, J., and G. Holzer. 1979. The photolytic degradation and oxidation of organic compounds under simulated Martian conditions. *Journal of Molecular Evolution* 14:153–60.

Oró, J., and A. P. Kimball. 1961. Synthesis of purines under possible primitive Earth conditions. I. Adenine from hydrogen cyanide. *Archives of Biochemistry and Biophysics* 94:217–27.

Oró, J., and A. P. Kimball. 1962. Synthesis of purines under possible primitive Earth conditions. II. Purine intermediates from hydrogen cyanide. *Archives of Biochemistry and Biophysics* 96:293–313.

Oró, J., and A. Lazcano. 1997. Comets and the origin and evolution of life. In *Comets and the origin and evolution of life,* ed. P. J. Thomas, C. F. Chyba, and C. P. Mckay, 3–27. New York: Springer.

Oró, J., and D. A. Rappoport. 1959. Formate metabolism by animal tissues II. The mechanism of formate oxidation. *Journal of Biological Chemistry* 243:1661–65.

Oró, J., A. P. Kimball, R. Fritz, and F. Master. 1959. Amino acid synthesis from formaldehyde and hydroxylamine. *Archives of Biochemistry and Biophysics* 18:115–30.

Oró, J., S. L. Miller, and A. Lazcano.1990. The origin and early evolution of life on Earth. *Annual Reviews of Earth and Planetary Sciences* 85:317–56.

Oró, J., T. Mills, and A. Lazcano. 1992a. Comets and the formation of biochemical compounds on the primitive Earth—A review. *Origins of Life* 21: 267–77.

Oró, J., S. W. Squyres, R. T. Reynolds, and T. M. Mills. 1992b. Europa: Prospects for an ocean and exobiological implications. In *Exobiology in Solar System exploration, NASA Special Publication 512,* ed. G. C. Carle, D. E. Schwartz, and J. L. Huntington, 103–25. Washington, D. C.: U. S. Printing Office.

Pizzarello, S., and J. R. Cronin. 1998. Alanine enantiomers in the Murchison meteorite. *Nature* 394:236.

Ponnamperuma, C. 1965. Abiological synthesis of some nucleic acid constituents. In *The origin of prebiological systems,* ed. S. W. Fox, 221–42. New York: Academic Press.

Ponnamperuma, C., ed. 1981. *Comets and the origin of life.* Dordrecht, The Netherlands: Reidel.

Schopf, J. W., ed. 1983. *Earth's earliest biosphere: Its origin and evolution.* Princeton, N.J.: Princeton University Press.

———. 1993. Microfossils of the Early Archean Apex chert: New evidence of the antiquity of life. *Science* 260:640–46.

———. 1999. *Cradle of life: The discovery of Earth's earliest fossils.* Princeton, N.J.: Princeton University Press.

Schrödinger, E. 1956. *What is life? and other scientific essays.* Garden City, N.J.: Doubleday.

Urey, H.C. 1952. The early chemical history of the Earth and the origin of life. *Proceedings of the National Academy of Sciences USA* 38:351.

Ward, P., and D. Brownlee. 2000. *Rare earth—Why complex life is uncommon in the Universe.* New York: Copernicus/Springer.

Wakamatsu, H., Y. Yamada, T. Saito, I. Kumasiro, and T. Takenishi. 1966. Synthesis of adenine by oligomerization of hydrogen cyanide. *Journal of Organic Chemistry* 31:2035–36.

CHAPTER 2

From Big Bang to Primordial Planet

Setting the Stage for the Origin of Life

ALAN W. SCHWARTZ AND SHERWOOD CHANG

INTRODUCTION

According to modern theory, life arose on the primitive Earth by a process of prebiotic chemical evolution. This process began with syntheses of organic chemical precursors of proteins, nucleic acids, and membranes in the early atmosphere and ocean, and ended with the emergence of life forms capable of self-replication—forms that could undergo Darwinian evolution through mutation and natural selection. Although most researchers accept this view of prebiotic evolution, opinions diverge on the nature and sequence of chemical events between the first and last stages. Many theoretical paths meander through this murky middle ground, but the track culminating in biological evolution remains unclear. While other chapters illuminate current progress in tracing this pathway, this chapter sets the stage, showing how life on Earth is related to—and a product of—a long sequence of cosmic events that preceded the formation of our planet.

Biochemistry must surely have grown out of geochemistry, that is, out of chemical reactions occurring among inorganic components derived from Earth's atmosphere, ocean, or crust. Even if we were to accept the tenets of **panspermia**—the idea that life was carried to Earth in meteorites from other bodies or in dust particles from interstellar space—we would still have to trace its beginnings to planetary processes elsewhere. And the nature of such processes is critical. Chemical evolution and planetary evolution are inextricably intertwined, so progress toward the origin of life may end at any stage if a host planet's physical or chemical development takes a wrong turn.

Since the early 1950s, when research into life's beginnings began in earnest, astronomers, astrophysicists, and cosmochemists have been making discoveries relevant to a general theory for the origin of life. Organic compounds in increasing numbers and complexity have been detected in the interstellar medium and comets. Many organic compounds of the sort that make up living systems—but unequivocally non-biological and extraterrestrial in origin—have been identified in **carbonaceous meteorites.** Spectroscopic observations in the Solar System have revealed organic compounds on **asteroids** as well as on the **giant planets** and their satellites. Minute grains of organic matter similar to those found in meteorites have also been detected throughout our galaxy and beyond. Clearly, organic matter occurs throughout the Universe as an integral component of cosmic evolution. And, with the detection of increasing numbers of planets around other stars, the prospect of identifying other life-harboring solar systems seems inevitable.

Throughout history, cultural concepts have shaped human perceptions of the Universe and humanity's place within it. Today, our very perceptions have changed, expanding beyond the reaches of the Solar System to the stars and interstellar clouds that populate limitless space. And our concepts have changed as well. Just as biological evolution implies that all organisms diverged from a common ancestor, cosmic evolution implies that all matter in the Solar System had a common origin. Life, we can now argue, is the product of countless changes in the form of primordial stellar matter, and these changes were wrought by astrophysical and cosmochemical, as well as geological and biological, evolutionary processes. In the context of cosmic evolution, the chain of events that led to life's beginnings extends back through planetary history to the origin of the Solar System, to the star-spawning turmoil of interstellar clouds, to the birth in stars of the biogenic elements that make up living systems. Indeed, the links in this chain constitute a cosmic history full of promise that life is widespread in the Universe.*

---

*Throughout this chapter, the reader is directed to images of astrophysical phenomena taken with the **Hubble Space Telescope** (HST). These pictures are made available by the Space Telescope Science Institute on the World Wide Web through links at http:// oposite.stsci.edu/pubinfo/SubjectT.html. To access a specific image, replace "SubjectT" in this web address with the bracketed code listed adjacent to the name of the phenomenon in the text. For example, the image referred to in the text as "Crab nebula [PR/2000/15/index.html]" can be retrieved at http://oposite.stsci.edu/pubinfo/PR/2000/15/index.html.

ORIGIN OF THE BIOGENIC ELEMENTS

Soon after the Big Bang, the expanding universe contained only three kinds of atoms—hydrogen (^{1}H), deuterium (^{2}H or D), and helium (^{3}He). As the swirling primordial cloud of atoms evolved, local concentrations of hydrogen and helium began to form. These denser cloud fragments became the birthplaces of the first generation of stars. Under the influence of gravitational forces, the cloud fragments contracted and released their potential energy as heat, some of which radiated to space while the rest raised the internal temperature of the forming stars. When these temperatures approached 10^7 degrees kelvin, atomic collisions became so energetic that **nuclear fusion** reactions occurred in the interiors of the stars. And when the rates of nuclear heat generation rose to balance the rates of heat loss, the stars stopped contracting and entered a stable period of stellar evolution fueled by the chemical reactions of nuclear fusion. During this quiescent period, their endowment of hydrogen was slowly "burned" and converted to helium. But once their stores of hydrogen were depleted, the stars contracted even more, and their temperatures rose again; at about 10^8 degrees kelvin, helium burning turned on, initiating formation of heavier chemical elements, particularly carbon (^{12}C) and two isotopes of oxygen, ^{16}O and ^{18}O.

When a massive star uses up its nuclear fuel and its central core contracts and collapses, gravity crushes the matter in its core to increasingly higher densities and internal temperatures reach billions of degrees. At these extreme conditions, nuclear reactions occur violently and the collapse is catastrophically reversed. A thermonuclear shockwave passes through the expanding stellar debris, fusing lighter elements into heavier ones and violently ejecting the material into the interstellar medium in a **supernova** explosion. In this way, first-generation supernovae released the initial inventory of ^{12}C, ^{16}O, and ^{18}O into the space between stars, the **interstellar medium,** leaving behind an exceedingly dense **neutron star.**

Other sources of carbon, nitrogen, and oxygen isotopes include the less energetic but more common novae, the stellar winds of red giant stars during their prime of life, and the planetary nebulae formed in the final stage of the evolution of **red giants,** when the last layers of the star blow away to reveal an underlying hot **dwarf star.** That mass loss plays an important role during the late stages of stellar evolution is revealed by the endpoints of this evolution—**white dwarfs** from planetary nebulae and neutron stars from supernovae—objects that are

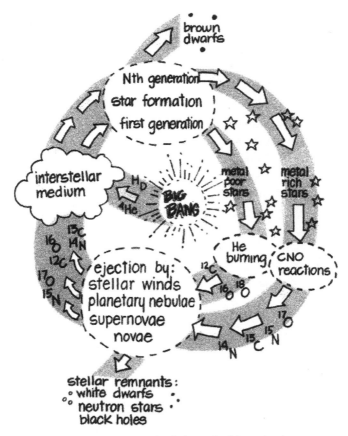

Figure 2.1. Over astronomical time, the biogenic elements formed by the processes in stars are cycled through the interstellar medium in many cycles of star formation, maturation, and death. (Cartoon by J. Wood, Smithsonian Astrophysical Observatory, Cambridge, MA.)

small, 2 solar masses or less, whereas mature stars reach 30 to 50 solar masses. The large difference in mass between main-sequence stars and the objects that remain after their demise represents the stellar material that is recycled into the interstellar medium.

The next generation of stars that formed incorporated the ^{12}C, ^{16}O, and ^{18}O their predecessors had ejected into the interstellar medium. These nuclides then acted as catalysts for more complex nuclear processing, during which hydrogen burning led to formation of ^{13}C, ^{14}N, and ^{17}O. Supernovae ejected the complete suite of C, N, and O isotopes into the interstellar medium accompanied by newly formed, even more massive chemical nuclides to serve as feedstock for subsequent generations of stars (see Crab nebula [PR/2000/15/index.html]). Through successive cycles of star birth and death, stellar **nucleosyntheses** produced all the other elements required for life (most notably, sulfur and phosphorus) and those needed for life's planetary habitats, and the elemental and isotopic composition of the interstellar medium continued to evolve over time. Thus, the atoms inherited by the Solar System have a complex genealogy traceable to innumerable stars in the past—stars in which these same atoms held temporary residence or in which they were formed from previously existing nuclides (figure 2.1).

Solar systems with rocky planets probably did not form until the heavy elements in their natal interstellar clouds exceeded some (as yet unknown) critical abundance levels. Just as the rates of star formation and death vary from region to region in the galaxy, so do the rates of growth of heavy nuclide abundances. These variations suggest that the timing and frequency of formation of solar systems that are like our own differed throughout the galaxy.

ORIGIN OF THE SOLAR SYSTEM

Interstellar Clouds

Today, the stars that make up most of the visible galaxy are embedded within a highly tenuous medium of gas and very fine dust. In regions where dust is thick enough to prevent the penetration of light, the interstellar medium appears as dark clouds against the stellar background. These clouds occupy about 5% of our galaxy's volume and are of various shapes and sizes. They can be as large as a **light-year** (9.5×10^{12} km) across, and they can contain as little as a few tens of solar masses or as much as millions of Suns. Their contents are mostly gas, predominantly hydrogen, but ~1% of their mass is present in the form of tiny dust grains approximately 0.1 micrometer (μm) across. The clouds are cold, having temperatures of the order 10 to 100 degrees kelvin, and are generally thin (10^4 molecules of hydrogen per cubic centimeter). Some have regions of higher gas and dust concentrations (of the order 10^5 to 10^6 molecules per cubic centimeter), called **cloud cores,** which serve as spawning grounds for stars (a good example is shown in Eagle nebula [PR/1995/44.html]).

The earliest intimate associations between minerals, organic compounds, and water occurred in interstellar clouds. Spectroscopic observations of such clouds in the vicinity of young stellar objects made

TABLE 2.1 COMPOSITION AND RELATIVE
ABUNDANCES OF INTERSTELLAR ICES AND
COMET ICES (H_2O = 100)

Species	Interstellar	Comets
H_2O	100	100
CO	2–25	5–30
CO_2	14–22	3–20
CH_4	0.5–4	1
CH_3OH	<3–18	0.3–5
H_2CO	<3–6	0.2–1
HCOOH	3–7	0.5
OCS	<0.2	0.5
NH_3	≤9–30	0.1–1.8
XCN^a	0.3–3.5	0.01–0.4

[a]CH-containing molecule in which X is unidentified.
Adapted from Gibb et al. (2000).

by the **Infrared Space Observatory** (ISO), launched in 1995, show the presence of ices of water and organic compounds mantling dust particles (table 2.1). Together with icy grains, **silicates** and **carbonaceous** particles make up the interstellar dust population, the last two representing condensates from the outflows of previous generations of stars.

Accompanying the dust in the gas phase of such clouds we find a striking variety of molecules, some of which are quite complex in structure whereas others, such as ethanol (C_2H_5OH), acetic acid (CH_3COOH), and formaldehyde (CH_2O), are fairly simple (table 2.2). Low temperatures permit the formation of unusual species (such as the cyanopolyyne, $HC_{11}N$) which are unstable under ordinary conditions. Irradiation by starlight and cosmic rays initiate ion–molecule reactions in the gas phase. Adsorption and recombination of atoms and free radicals occur on grain surfaces. These low-temperature processes account for the bulk of the organic chemical reactions that occur in quiescent interstellar clouds (figure 2.2). In regions of star formation and in the vicinity of energetic stellar phenomena such as supernovae, energy fluxes are more intense, temperatures are higher, and the relatively more complex molecules are present.

Solar Nebula

Observations of regions of star formation by use of modern telescopes (particularly the Hubble Space Telescope and ISO), combined with developments in astrophysical theory, form the basis for an emerging picture

TABLE 2.2 ELEMENTAL COMPOSITIONS OF
INTERSTELLAR MOLECULES IN THE GAS PHASE

C, H, O, N	C, H, N	C, H, O	C, H	C, N	C, O
HNCO	HCN	HCO	CH	CN	CO
H_2NCHO	HNC	HCO^+	CH^+	C_3N	CO_2
	HCNH	$HOCO^+$	CH_2		C_3O
	H_2CN	H_2CO	CH_4		
	H_2CNH_2	H_2COH^+	C_2H		
	H_3CNH_2	H_2COH	C_2H_2		
	HCCN	HCOOH	C_2H_4		
	H_2CCN	H_2CCO	C_3H		
	H_3CCN	H_3CCHO	C_3H_2		
	H_2NCN	H_3CCH_2OH	CH_3C_2H		
	HC_3N	H_3COCH_3	C_4H		
	HCCNC	HC_3HO	C_4H_2		
	HNCCC	H_3CCOCH_3	C_5H		
	HC_3NH^+	$HCOOCH_3$	CH_3C_4H		
	H_2CCHCN	H_3CCOOH	C_6H		
	H_3CCH_2CN		C_7H		
	HC_5N		C_8H		
	H_3CC_3N		PAH		
	HC_7N				
	H_3CC_4CN				
	HC_9N				
	$HC_{11}N$				

PAH denotes polycyclic aromatic hydrocarbons; chemical species detected but not listed here include CS, CP, NP, NH, NH_2, NH_3, H_2O, H_2S, HCS^+, OCS, SO_2, C_2S, C_3S, HNCS, and CH_3SH.

of Solar System formation. Although much about the process is incompletely understood, the main stages have been postulated, quantitative theoretical models are being developed for the major steps involved, and observations are being used to test the models.

The relatively high density in the cores of interstellar clouds is thought to have arisen either from a diffusive separation of charged and neutral species in the cloud's magnetic field or from a violent disturbance nearby, such as a supernova shock wave (which may have triggered the formation of the Solar System). Regardless of cause, the elevated cloud density led to a state of gravitational instability in which the forces (turbulent motions, thermal pressure, and the magnetic field) that normally prevent core contraction failed to withstand the force of gravitation, so that the core dynamically collapsed. During this process, gravitational energy was released and radiated away, efficiently at first, and the cloud fragment's angular momentum was translated into rotational motion. Since **centrifugal forces** do not affect gravity in directions parallel to the

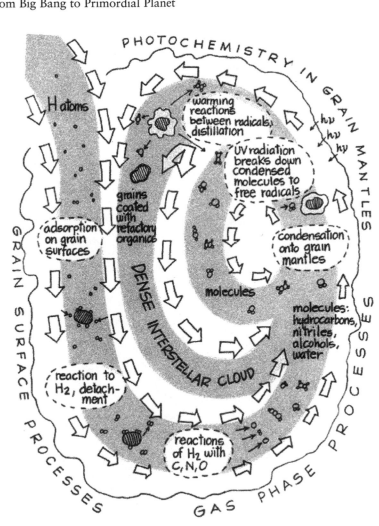

Figure 2.2. Gases and grains are subjected to a variety of physical and chemical processes during their residence in interstellar clouds. (Cartoon by J. Wood, Smithsonian Astrophysical Observatory, Cambridge, MA.)

rotational axis, the cloud core collapsed to a disk rather than to a sphere. Its density increased first to 10^{10} and then to 10^{20} molecules per cubic centimeter as the cloud diameter decreased to less than 1,000 **astronomical units** (AU; the mean distance from the Earth to the Sun) and flattened into a rotating disk. The magnetic field inherited from the cloud fragment grew as the disk spun down, and an initially rapid influx of gas to the center of the disk produced a **proto-Sun,** which continued to

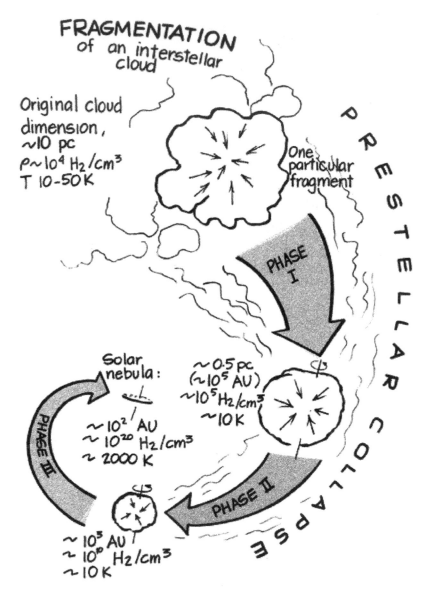

Figure 2.3. After an interstellar cloud fragments, it goes through several phases of prestellar collapse and evolution. (Cartoon by J. Wood, Smithsonian Astrophysical Observatory, Cambridge, MA.)

grow by inward migration of matter within the disk in response to the outward transport of angular momentum. We call this rotating circumstellar disk, with its embedded protostar, the **solar nebula** (figure 2.2). Recently, such disks have been observed directly in star-forming regions (see, for example, Orion nebula [PR/1995/45.html] and Proplyds [PR/1994/24.html]).

During the formation of the Solar System, the thermal state of the nebula was determined by a balance between the gravitational energy released by material accreting onto the proto-Sun and the energy delivered to the solar nebula and radiated away from its outer surfaces. Initially, the collapsing cloud fragment remained cold (figure 2.3); but as material rapidly accreted onto the embryonic Sun, the density of the solar nebula increased until it became opaque to the thermal radiation of the proto-Sun, trapping heat energy and causing the temperature at the center of the disk to increase rapidly. Close to the proto-Sun, temperatures became high enough to melt silicate minerals, while at the outer reaches of the nebula they remained low enough to allow even such volatile gases as carbon monoxide (CO), nitrogen (N_2), and methane (CH_4) to condense into ices. Energetic processes operating near the accreting proto-Sun combined to drive a flow of gas, a **solar wind**, that accelerated outward along the rotational axis.

As the rate of accretion onto the proto-Sun dwindled, the nebula slowly cooled while preserving a thermal gradient in which temperatures decreased with increasing distance from the forming star. On the basis of the range of ages inferred for astronomically observed young stars having circumstellar disks, the period from the collapse of the interstellar cloud to the development of a cool solar nebula is estimated to have spanned several to perhaps ten million years. (Interestingly, this model of the origin and development of the solar nebula, though continuously upgraded and refined on the basis of new observations, is rooted firmly in theories first articulated by Immanuel Kant in 1755 and Pierre-Simon Laplace in 1796.)

Throughout these stages of solar nebula evolution, the fate of in-falling interstellar matter—its transformation and ultimate redistribution among bodies of the Solar System—was governed by complex interactions among physical, chemical, and dynamic processes (figure 2.4). These included thermochemical and photochemical reactions, irradiation by high-energy ions and electrons, vaporization, condensation, fluid transport, dispersion by collisions, and ejection back into the interstellar medium. Because many of these processes were driven by the proto-Sun,

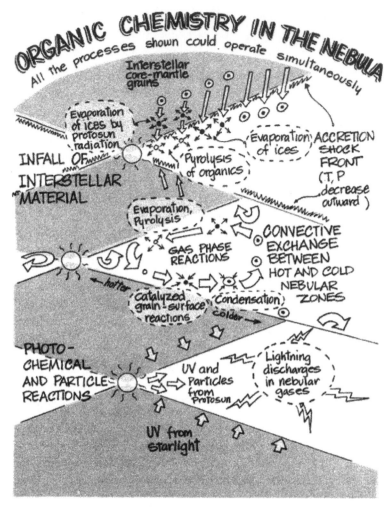

Figure 2.4. Gases and solids in the solar nebula undergo a wide variety of transformations as the result of physical, chemical, and dynamical processes operating in the solar nebula as it evolves. (Cartoon by J. Wood, Smithsonian Astrophysical Observatory, Cambridge, MA.)

their effect was to create a highly energetic environment whose intensity diminished with increasing distance from the forming Sun.

Planets

As the solar nebula evolved, clumps of gas and dust coagulated to form the seeds from which planets formed (see AB aurigae [PR/1999/21]). Ice and mineral grains that flowed into the hot region near the proto-Sun were vaporized, and some of the vapor was transported outward to cooler regions where it recondensed into new grains. These submicron-sized solar nebula condensates, together with grains inherited intact from the cloud core, coalesced over time to form objects centimeters broad. Because of the thermal gradient across the nebula, the objects in the inner region were composed primarily of refractory inorganic minerals, whereas those in the outer region were predominantly made up of water and other ices mixed with organic compounds. Between these regions, objects that contained large proportions of both ices and minerals occupied a transition zone. All of these particles eventually settled to the midplane of the nebula, forming a thin disk densely populated by solid objects; there, because of their crossing orbits and low collisional velocities (<10 m/sec), they grew in size as they collided with one another and stuck together. According to computational simulations, the aggregation of solid grains into meter-sized boulders would have taken less than a million years, and the accretion of planetesimals—bodies meters to tens of kilometers across—from 100,000 to 10 million years.

When kilometer-size planetesimals pass close to each other, their orbits can be perturbed by mutual gravitational attraction; collisions can produce either fragmentation or further growth. Simulations show that the maximum mass in populations of such objects increases steeply as the smaller planetesimals are fragmented and aggregated onto the larger ones, forming a sort of "feeding zone" where runaway accretion can occur. During this stage, a few bodies—the **protoplanets**—grow much more rapidly than the others. When this happens, most of the mass of the population is concentrated in a few large objects, so collisions between them can be cataclysmic. As the feeding zones of the protoplanets are swept clear of debris, gaps form in the circumstellar disk and runaway accretion stops (see Disk Gap [PR/1999/03]). Simulations of these processes designed to mimic settings like those of our own proto-Solar System all seem to result in formation of two to five terrestrial planets in the inner reaches of the system (where Mercury, Venus, Earth, and Mars now reside) over an accretion time of about 100 million years.

While runaway accretion was forming the inner planets of the Solar System, the same process in the outer parts of the solar nebula led to aggregation of the huge rock–ice core of Jupiter, some 15 to 20 times larger than the mass of the Earth. Ever increasing amounts of hydrogen and helium were held by gravitation in Jupiter's atmosphere as the planet's massive core grew larger. Studies show that accretion of gas from the nebula was initially negligible; but once the core attained a critical mass, gas accreted at an increasingly rapid rate, eventually greatly exceeding the rate of mass increase from planetesimal in-fall. Similar processes are likely to have formed Saturn, Uranus, and Neptune; but for these planets, the sizes of the cores formed and the amounts of gas accreted were smaller than those of Jupiter.

Exactly what terminated the accretion of gases on the giant planets is uncertain. By comparison with the elemental abundances in the Sun, none of these planets, including Jupiter, seems to have acquired as large a complement of solar nebular gas as would be expected. Possibly, the gaseous nebula was already beginning to dissipate by the time the planets' cores attained critical mass. But the mechanisms of such dissipation are not fully understood, so other explanatory processes, such as photo-evaporation by intense radiation from nearby young stars or the onset of an extremely intense solar wind, have also been suggested (see Evaporating Globules [PR/95/1995/44.html]). Astronomical observations of young stars show that the initial circumstellar dust cloud has already been cleared from around those older than about 10 million years. This means that for these stellar nebulae, and perhaps for our own solar nebula, planetary growth primarily occurred after the unconsolidated gas and dust of a circumstellar disk had been dissipated.

Comets, Asteroids, and Meteorites

The growth of the giant protoplanets had consequences elsewhere, beyond the regions of the Solar System relatively nearby. Many of the ice-rich planetesimals that accreted in the outer solar nebula were gravitationally scattered into eccentric orbits. Some were ejected from the Solar System; others were injected into the inner Solar System, where they were destined to collide with the terrestrial planets or the Sun; still others were bound into orbits that extended beyond 10,000 AU to make up the **Oort cloud,** the spheroidal shell of bodies that encircles the Solar System. Over time, gravitational influences from outside of the Solar System—a result, for example, of passing stars or interstellar clouds—perturbed the orbits of the Oort cloud objects, sending some back into the Solar System as long-period comets. As these bodies approach the Sun, they are gradually heated, and their ices sublime to release dust and gas into their tail-like comas, a phenomenon that makes them visible from Earth (see, for example, comet Hale-Bopp [PR/1995/41.html]).

In the last few years, another reservoir of icy bodies has been discovered. These objects, whose orbits lie beyond that of Neptune and mostly beyond that of Pluto, are members of the **Kuiper belt** group, which is thought to be the source of **short-period comets**. The ISO has been used to identify the ices that coat the surfaces of several of these distant objects. These small icy bodies, formed and preserved at and beyond the outer edges of the Solar System, are prime candidates as storehouses of intact interstellar matter.

Recent astronomical observations of comets approaching the Sun, including Hyakutake and Hale-Bopp, have revealed the presence of a rich variety of simple organic compounds (table 2.3) chemically related to those observed in dense interstellar clouds. Analysis of carbonaceous dust in comet Halley's coma revealed elemental compositions compatible with mixtures of small interstellar molecules or their polymers (table 2.2). The

TABLE 2.3 COMPOUNDS OBSERVED IN
THE COMAS IN COMETS ($H_2O = 100$)

Compound	In Hale-Bopp	In other comets
H_2O	100	100
CO	20	20–1
CO_2	20	20–3
H_2CO	0.1	1–0.1
CH_3OH	2	7–1
$HCOOH$	~0.05	
$HNCO$	0.1	
NH_2CHO	~0.01	
$HCOOCH_3$	~0.05	
CH_4	~0.6	
C_2H_2	~0.1	
C_2H_6*	~0.3	
NH_3	0.6	
HCN	0.2	0.2–0.05
HNC	0.04	
CH_3CN	0.02	
HC_3N	0.03	
H_2S	1.5	1.5–0.2
H_2CS	~0.02	
CS	0.2	0.2
OCS	0.5	
SO	~0.5	
SO_2	0.1	
S_2*		0.005

*C_2H_6 and S_2 have not yet been seen in interstellar clouds.
Adapted from Irvine et al. (2000).

compositions and relative abundances of the ices thought to comprise the greater part of the nuclei of comets can be inferred from measurements of the gases in their comas. These reconstructions show a remarkable resemblance to the composition and relative abundances of ices recently measured in interstellar clouds (table 2.1). The ices present both in long-period (Oort cloud) and short-period (Kuiper belt) comets appear to be quite similar. These and other findings support the notion that, in large measure, the icy interstellar grains and associated organic matter survived the transition from interstellar cloud to solar nebula, preserved in icy plan-etesimals that became the building blocks of comets.

As the Solar System formed, gravitational scattering by Jupiter not only influenced the orbits of comets but also disrupted the formation of a planet in the **asteroidal belt,** the zone of space between the orbits of Jupiter and Mars. Asteroidal bodies were scattered by gravitational perturbations, depleting the zone of its population of planetesimals as many were propelled into the inner reaches of the Solar System. Although planet formation in the asteroidal belt was thus frustrated, we can study the objects preserved there; eventually, such studies should provide a sound basis for confirming or rejecting current models of the earliest stages of planet building.

Although subject to a logarithmic decline after an early period of es-pecially intense collisional interactions, impacts in the asteroidal belt have sent large and small fragments of their parent bodies into Earth-crossing orbits throughout Solar System history. Similarly, the passage of comets has deposited trails of cometary dust through the inner Solar System. Today, some of this dust enters Earth's atmosphere, where it decelerates gently and is captured in collectors deployed by high-flying aircraft. These small comet grains (≤ 50 μm), together with asteroidal fragments of similar size, comprise the **interplanetary dust particles** (IDPs). Laboratory studies of these objects afford a tantalizing glimpse of the composition and history of objects formed elsewhere in the Solar System. However interesting to researchers, IDPs are too small to per-mit the comprehensive organic analyses that can be carried out on larger impacting bodies—meteorites.

Meteors are chunks of debris left over from the formation of the Solar System. When we see them coursing through the sky, we call them "shooting stars"; but once they hit the planet's surface, we call them **meteorites.** Many meteorites are metallic, containing shiny grains of an iron–nickel alloy like that in the interior of Earth. But others are min-eralic masses, evidently dislodged by violent collisions from the surfaces of rocky asteroids, and even the Moon and Mars. Among these stony objects are the **carbonaceous chondrites,** so named because they contain a speckling of tiny glassy balls (chondrules) and up to 5% of the element carbon, largely in the form of dark "coaly" carbonaceous organic matter. This group of meteorites is of special interest because their organic chemical contents include many compounds found in biochemistry even though they originated by purely nonbiological processes.

The most famous carbonaceous chondrite studied to date fell on September 28, 1969 near Murchison, Australia. Known as the **Murchison meteorite** (or, simply, "Murchison"), this meteorite is an organic-rich object. During the past two decades, Murchison's organic constituents have been studied in detail, as have been those in carbonaceous meteorites retrieved from the Antarctic ice sheet. Prior to 1969, studies of the organic chemistry of such meteorites had been confounded by the presence of ubiquitous and all too often obscuring terrestrial contamination. But research techniques have improved since then. Recent analyses, particularly of Murchison, show that the organic compounds detected are unquestionably extraterrestrial, and that they comprise a remarkable concentration and diversity of organic compounds embedded within a clay and carbonate mineral matrix.

Life requires both water and a diversity of organic chemical building blocks. Hence it is noteworthy that the presence of liquid water was a prerequisite for the formation of the mineral assemblage found in such meteorites. Moreover, among the many classes of organic compounds detected, these meteorites contain amino acids, fatty acids, and both purines and pyrimidines—all building blocks of life (table 2.4). In all classes of compounds studied, homologous series occur, in which the molecular abundances decrease logarithmically with increasing carbon number. Carboxylic acids up to 12 carbons in length have been detected, and nearly all possible structural isomers of compounds having low carbon numbers have been found. But among those classes of compounds relevant to biology, only a very small fraction of the meteoritic isomers are capable of playing a biochemical role. However, amino acids in the Murchison have recently been shown to contain a slight excess of the levorotatory L-**enantiomer.** (Some astrophysicists argue that this enantiomeric excess was induced in the interstellar medium by circularly polarized light arising from dust-scattering in regions of high-mass star formation. But regardless of how it was produced, this **nonracemic** amino acid mix appears to provide the first evidence of a naturally occurring chiral process unrelated to biology.)

TABLE 2.4 ORGANIC COMPOUNDS
IDENTIFIED IN METEORITES

Compound	In meteorites	In biology
Biochemical building blocks		
Amino acids	+[a,b]	Proteins
Fatty acids	+[c]	Membranes
Glycerol	+	Membranes
Inorganic phosphate	+	Membranes and nucleic acids
Purines	+	Nucleic acids
Pyrimidines	+	Nucleic acids
Ribose and deoxyribose	—	Nucleic acids
Other biochemical components		
Alcohols	+	+
Aldehydes	+	+
Amides	+	+
Amines	+	+
Mono- and dicarboxylic acids	+	+
Hydroxy carboxylic acids	+	+
Aliphatic hydrocarbons	+	—
Aromatic hydrocarbons	+	—
Ketones	+	—
Phosphonic acids	+	—
Sulfonic acids	+	—
Sulfides	+	—

[a]At least one of the isomers identified in meteorites occurs in living systems;—, not detected.
[b]Most meteoritic amino acids are not found in living systems.
[c]None of the short-chain meteoritic acids occurs in biological membranes.

The critical connection on Earth between liquid water and life, with its organic composition, lends special significance to the evidence for liquid water and organic compounds in carbonaceous meteorites. Apparently, prebiotic evolution of some kind occurred early on parent bodies in the asteroidal belt, but fell short of the origin of life itself. Nitrogen, hydrogen, and sulfur measured in meteoritic organic matter are anomalously enriched in their heavier isotopes (figure 2.5), a finding that suggests an interstellar origin for most, if not all, of the organic matter detected. Only in interstellar molecules are more extreme ratios of deuterium to hydrogen (D/H) known to occur, in some cases approaching values of 0.3 (the D/H ratio in galactic hydrogen is ~2 × 10^{-5} and that for ocean water is ~10^{-4}). Tiny grains of graphite and

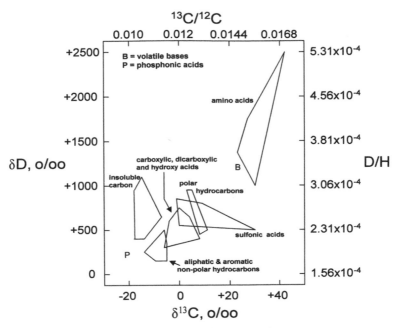

Figure 2.5. The organic components of the Murchison meteorite are composed of isotopes of carbon (^{12}C and ^{13}C) and hydrogen (^{1}H and D, deuterium) that vary widely and lie well outside the ranges of values found in organic matter on Earth.

diamond in Murchison also exhibit isotopic properties that signify origins as stardust ejected from earlier generations of carbon stars. And the spectral properties of carbonaceous meteorites resemble those of the abundant C-type main-belt asteroids that occur predominantly near the center and outward to the distant edge of the asteroidal belt.

All these considerations lead to the conclusion that parent bodies of at least some carbonaceous meteorites formed by co-accretion of ice-rich and rock-rich planetesimals in the transition zone between the interstellar cloud and the solar nebula. During subsequent internal heating, ices melted, and the resulting liquid water transformed anhydrous minerals to phases such as clays and carbonates. **Hydrolysis** and other aqueous reactions altered interstellar precursors (table 2.2) to yield some of the organic compounds listed in table 2.4. Conversion of nitriles to organic acids is an example.

Small bodies continued to fall onto proto-planets throughout the accretionary process. Indeed, such impacts evidently occurred, though with exponentially decreasing frequency, over the first 700 million years of Solar System history, after which a low, more or less steady state of in-fall became the norm, a situation that has maintained up to the present. The record of early impacts is preserved in the cratered surfaces of the Moon, Mercury, and Mars, and the ages of cratering events on the Moon have been established from rock samples brought to Earth by NASA's Apollo Project. As discussed in chapter 6, age measurements on such samples show that large impactors struck the Moon as late as ~3,800 million years ago. The prediction that the final stages of planetary formation involve collisions between large bodies of similar size is consistent with the idea that the Moon was formed by material ejected into orbit by a Mars-sized object (~6,700 km in diameter) that struck the Earth. The occurrence of such giant collisions also explains variations in orbital inclinations, rotational periods, and tilted rotational axes among the various inner planets.

EARTH

Early Geophysical Evolution

According to the current scenario of planetary origin, Earth formed in a violent process over a period of about 100 million years. Impacting bodies deposited their **kinetic energy** at depth as well as at the surface; and as the hot planetary embryo grew, iron separated from silicate to form a metallic core. Collision with another large embryonic planet tilted Earth's rotational axis and ejected a portion of its mantle into orbit where it cooled and coalesced to form the Moon. Collisions continued at a declining rate as Earth captured the remaining bodies in its feeding zone. The newborn planet developed a magnetic field, a highly convective mantle, and a molten surface. As the input of energy from accretion diminished over time, surface temperature dropped, a surface scum of rock solidified, and water rained out to form an ocean. Magnetic field and atmosphere provided shielding against highly energetic and intense radiation from the young Sun. Surface temperatures at or below 100°C prevailed as early as 4,400 million years ago.

Collisions continued over the first many hundred million years of Earth's history, perhaps occasionally with huge impactors, but increasingly with ever smaller bolides that yielded correspondingly weaker effects and faster recovery times. In impacts of Earth with bodies ~450 km in diameter, ejected hot rock vapor encircled the globe and its heat radiated downward, evaporating the early ocean to produce a dense steam atmosphere having a pressure of perhaps several hundred **bars**. With continued radiative heating, the exposed surface melted to considerable depth. Eventually, radiation of the heat to space cooled the atmosphere and surface. Once the environment had cooled to the **critical point** of water, precipitation began and a new ocean rained out, a return to the pre-giant impact environment that appears to have been remarkably rapid, calculated to have taken only a few thousand years.

Devastating, ocean-evaporating impacts precluded the survival of emergent ecosystems. But the short duration of their effects and their strongly diminishing frequency over time assured the eventual sustainability of life on the surface. Various workers theorize that the last truly major impact occurred either as early as 4,400 million years ago or as late as 3,800 million years ago. Undoubtedly, impacts constrained the timing of the origin of life, but such large uncertainties in the calculations leave open the earliest time at which life could have gained a permanent foothold on the planet. Little is known about how long it takes for the origin of living systems. If the required time scale is short, life may have arisen more than once between major impacts (and, of course, the hazards to life posed by such impacts decrease with depth from surface environments to the ocean deeps where living systems might have found a safe haven from the devastation at the surface).

Although the Earth is 4,550 million years old, a substantial readable rock record, direct evidence of Earth's early history, dates from ~3,900 million years ago. Nonetheless, the earliest 650 million years must have been a time of extensive geophysical change. The dwindling size and frequency of impacts lengthened the time between major environmental perturbations. Heat flow from Earth's interior was initially very high, permitting dissipation of thermal energy left over from accretion. But by about 3,800 million years ago, heat flow had evidently diminished to a level only about four times greater than that at present, and both volcanism and hydrothermal activity had similarly lessened. As the new-formed Moon gradually receded from the Earth, day length increased, and the amplitude of ocean tides decreased. The total luminosity of the young Sun brightened to about 70% of its current level, while its high early ultraviolet output decreased substantially from levels that were initially more than 30 times as intense as now. The crust began to sort itself into continental and oceanic realms, the atmosphere and oceans formed, and land surface grew in areal extent.

Atmosphere–Ocean System

Studies of the elemental and isotopic compositions of **noble gases** in samples derived from Earth's mantle indicate that these gases were being released into the atmosphere as early as 4,400 million years ago. Since nitrogen is relatively unreactive and behaves like a noble gas, it, too, outgassed early on. Dissolved in hot surface rocks or trapped in sediments and then cycled between the surface and the interior, water and such carbon-containing gases as carbon dioxide, carbon monoxide, and methane came to be partitioned between the mantle and the ocean–atmosphere system. The presence of a primordial isotope of helium (^{3}He) in mantle rocks suggests, however, that Earth may have gravitationally captured gases from the solar nebula before the nebular cloud completely dissipated. If so, most primordial gases appear to have been lost since then, and if Earth's Moon was actually formed by a giant impact, this devastating event could have provided the mechanism to blow away Earth's initial hydrogen-rich (solar-composition) atmosphere.

After loss of its original, solar nebula–derived gaseous envelope, Earth's atmosphere was replenished. Model studies of the solar nebula show that where Earth's planetesimal building blocks formed, temperatures were too high to permit condensation and accumulation of water and other volatile materials. This means that the water and gases making up Earth's secondary atmosphere must have been imported from more distant regions of the Solar System, most likely by comets and asteroids that impacted the planet after the close of the major epoch of planetary formation. The size of the atmosphere–ocean system at any particular time during this stage of development depended on the net balance between gains and losses of volatile materials resulting from impacts and from geologic cycling between mantle and surface. As a consequence of the decline in the rate of impacts during the earliest several hundred million years of the planet's history, it is likely that the bulk of the ocean–atmosphere system accumulated relatively early, rather than later, during the period between 4,400 and 3,800 million years ago.

According to solar evolution models, the luminosity of the young Sun was ~30% lower than it is today. Earth's surface temperature would have been cooler, and a global "greenhouse" would have been necessary to keep the early ocean from freezing. The three **greenhouse gases** most likely to have been available were ammonia (NH_3), methane (CH_4), and carbon dioxide (CO_2). Of these, ammonia and methane are more efficient in promoting the greenhouse effect, but both are subject to rapid photodecomposition and could have played important roles only if their supplies were being rapidly replenished. No sizable and thermodynamically plausible source of ammonia is known, and it remains to be demonstrated whether hydrothermal generation of methane on the early Earth would have been adequate to achieve the needed greenhouse effect. Although less efficient as a greenhouse gas, carbon dioxide seems certainly to have been the dominant carbon-containing gas released by volcanism and hydrothermal activity on the ancient planet, just as it is today. Moreover, vaporization of impactors striking the planet would have also released carbon dioxide (along with carbon monoxide) into the atmosphere, and in much larger amounts than methane. Calculations modeling the climate of the early Earth show that the presence of atmospheric carbon dioxide in amounts ranging from 100 to 1,000 times the amount in the present atmosphere would have been adequate to maintain the required greenhouse. Thus, the lack of heat generated by the low-luminosity young Sun was offset by an atmospheric greenhouse effect, with current considerations favoring assignment of most of this effect to an atmosphere rich in carbon dioxide and containing only small amounts of carbon monoxide and methane.

In contrast with this picture of a carbon dioxide–rich atmosphere, most scenarios for chemical evolution call for a highly reducing atmosphere—one rich in methane rather than carbon dioxide—in which to synthesize prebiotic organic compounds. This expectation is based on the success of S. L. Miller's pioneering experiments of 1953 in producing amino acids by passing an electric discharge through a gaseous mixture of methane, ammonia, and water vapor (experiments well summarized in chapter 3).

Even before Miller's work, however, the proposed existence of a highly reduced methane-rich primordial atmosphere was a controversial issue, and it remains so today. The chemical composition of gases released from Earth's interior depends on the equilibrium oxidation state and temperature of the source regions from which they are emitted. If metallic iron is present in such regions, as during the major epoch of Earth's accretion and core formation, carbon monoxide (CO), rather than methane or carbon dioxide, is the predominant carbon-containing gas released. Carbon monoxide is also the primary form in which carbon occurs in a primordial solar nebula and in gas pulses generated by vaporization of impactors that are metal-rich. And geologic processes commonly produce carbon dioxide—after metallic iron separated from Earth's mantle to sink to its core, carbon dioxide was the dominant carbon-containing gas released, both at high temperatures by volcanism and at lower temperatures in hydrothermal systems. In none of these situations is methane other than a minor component.

Unfortunately, however, the answer to this important question of the carbon source on the early Earth remains elusive. Geologic recycling of the Earth's crust has left no remnant of the earliest 650 million years of Earth's history from which the composition of the atmosphere could be read, so the dominant form in which carbon occurred in the early environment—whether as CO, CO_2, or CH_4—remains a matter of conjecture. It is only when the truncated geologic record begins, at 3,800 million years ago, that we have firm evidence to resolve the matter. But by this time, there is no question: Igneous rocks of this age have original oxidation states that show carbon dioxide to have been the predominant carbon gas issuing from Earth's interior.

Though severely heated and pressure-cooked, metasedimentary rocks from near Isua in southwestern Greenland provide the earliest and widest window into what Earth's environment was like 3,800 million years ago. The rocks from this region are a mixture of volcanic debris and water-laid sediments, including lavas that congealed at high temperatures as well as ordinary (if now geochemically altered) deposits containing carbonate minerals and particles of graphitic carbon. The chemistry of these rocks shows that carbon dioxide (and presumably nitrogen as well as various noble gases) was present in the atmosphere, and suggest that there was higher heat flow from Earth's interior and, presumably, more active volcanism and hydrothermal activity than today. The sediments of the sequence signify that surface temperatures were below 100°C, and their very presence—even in this small preserved sliver of ancient terrain—suggests that there were extensive bodies of liquid water. Weathering, a hydrologic cycle, and a carbon geochemical cycle—all perhaps driven by a form of geologic tectonism akin to that occurring in island arc systems of today—can also be inferred. To a considerable extent, the Isua rocks suggest surroundings similar to those of the present day, though the environment probably lacked atmospheric molecular oxygen, a byproduct of green plant photosynthesis. In general, the setting they suggest seems to differ little from the shallow marine environment shown by the 3,500-million-year-old sediments of Western Australia discussed in chapter 6—the source of the earliest compelling evidence of life now known (and, like the Isua setting, a wave-washed volcanic platform subject to episodic volcanism and hydrothermal activity). It is even conceivable that such environments may have existed as early as 4400 million years ago, although because the early rock record is missing, there is no way to tell for sure.

Prebiotic Organic Syntheses

If carbon dioxide rather than methane was the dominant gaseous form of carbon in the early environment, synthesis in the atmosphere of organic compounds needed for the origin of life would have been decidedly more difficult. This comparison is backed by many laboratory experiments, and the conclusion holds regardless of whether the reacting gases are irradiated by electrons, photons, or ions (energy sources selected, respectively, to simulate the effects of electrical and coronal discharges, solar ultraviolet radiation, and both cosmic rays and magnetospheric particles). It also applies to gas mixtures that span oxidation states from reducing (CH_4–N_2–H_2O) to neutral (CO_2–N_2–H_2O). The overall productivity of the reactions studied depends strongly on the "reducing power" of the gas mixture, the proportion of hydrogen present in the gas mixture. Reactions in which methane is used as a starting material yield abundant organic compounds of considerable structural diversity, whereas those using carbon dioxide produce very low yields and a limited array of compounds.

Although the actual nature of the early atmosphere remains to be determined, the experimentally well established difficulty in synthesizing organic compounds under chemically neutral conditions has led researchers in this field to expand their efforts in two new directions— studies of organic syntheses in hydrothermal systems (such as those associated with deep-sea fumaroles) and studies of the importation of extraterrestrial organics to the primordial Earth via comets and asteroids.

In several important respects, the ocean–crust interface in hydrothermal systems is similar to the ocean–atmosphere interface. Inasmuch as life itself must have emerged as a phase-bounded system, the formation, dissipation, and reformation of small-scale interfaces must have been a prerequisite for life. Relevant small-scale structures are provided near the Earth's surface by aerosols, organic films, and volcanic and interplanetary dust particles; and, at depth in the world's oceans, by hydrothermal minerals, chemical precipitates, and vesicle-like structures of organic or mineralic composition. Energy sources requisite for production of organic compounds occur in both environments—at the planetary surface, primarily sunlight; at oceanic depths, chemical disequilibria between hydrothermal fluids and seawater. Both interfaces can act as collecting zones for organic compounds formed or carried in from above or below. But agencies that cause decomposition of organics also occur in both—in near-surface settings, photodecomposition in the atmosphere or shallow seas and geothermal heating of products buried in sediments; at oceanic depths, the recycling of synthesized organics into fumaroles or the Earth's hot interior.

Hydrothermal systems are present today both in shallow near-shore environments and in deep marine environments. With regard to prebiotic syntheses on the early Earth, however, deep oceanic settings would have afforded advantages not offered by those of the near-surface: They would have been far less susceptible to effects of planetary impacts even while maintaining an intimate connection to the Earth's prime source of chemical reducing power—oceanic, dissolved, ferrous iron. Thus, deep-sea fumarolic vents are among the most promising of all early environments. There prebiotic evolution could have been sustained over long periods, with minimal loss of the chemical reducing power necessary for nonbiological synthesis of organic compounds.

Experimental evidence is emerging in support of the view that organic compounds can be formed under conditions prevalent in hydrothermal systems, but whether these dynamic systems were actually the spawning grounds for life on Earth remains an open question. Interestingly, **hyperthermophilic** microbes, which thrive in hydrothermal environments, are among the present-day organisms exhibiting the most ancient ancestral lineages, as determined by ribosomal RNA sequences. As discussed in chapter 6, it is not known whether this association of primitiveness and hyperthermophily reflects the origin of life in such settings or the selective preservation in them of microbial lineages following catastrophic destruction of surface ecosystems. In either case, however, conclusions based on phylogeny seem premature, inasmuch as countless unknown organisms remain to be characterized in the biosphere. The prebiotic Earth did not lack environments where the origin of life might have occurred. The problem is to understand which of these was most suitable and to define the steps taken during chemical evolution.

Although many of the molecules regarded as potential starting ingredients of life occur in comets and parent bodies of carbonaceous meteorites, it is not clear that they could have contributed to chemical evolution and the origin of life. There are at least two reasons for this uncertainty. First, the proportion of organic-rich impactors that would have reached the planet's surface intact is highly uncertain. Some impactors do reach Earth even today, as shown by Murchison and other meteorites, but we can only speculate that the organic matter they supplied was adequate for prebiotic evolution. Second, we do not know whether, in the event of globally catastrophic impacts, surface conditions on the primitive planet would have allowed the survival of imported organics, conditions that could well have destroyed native organics as well. Sustained prebiotic evolution, like the ecosystems it eventually spawned, could not have occurred until major impacts no longer threatened the environment.

Theoretical assessment of extraterrestrial sources of organics that may have played a role in life's beginnings points to IDPs as the likely predominant carriers. Today, such particles bring to Earth about 4000 million grams of carbon per year, and they may have contributed much larger amounts early in the planet's history. These small particles dominate the mass flux of extraterrestrial materials arriving at Earth's surface, even though much larger, even boulder-sized, meteorites also rain in. Survival of intact organics in the larger bodies is problematic because the energy released during hypervelocity collisions with Earth's surface is sufficient to vaporize much of the projectile (as well as a comparable amount of the impacted surface). It is conceivable that some material in large impactors could survive, but how much is unknown and has yet to be modeled quantitatively. And many unknowns attend the question of IDPs: What organic compounds do they actually contain, other than aromatic hydrocarbons? Could such compounds survive the heating that accompanies the deceleration of such particles as they plunge through the atmosphere? Were the amounts and types of the surviving organics sufficient to generate biological systems by chemical evolution as they dissolved in, and were diluted by, Earth's oceans? With all the uncertainties involved, the only firm conclusion we can draw is that both native organics and extraterrestrial organic matter were probably present in the prebiotic milieu (figure 2.6).

Wherever the organic building blocks of the first biota came from, their mere synthesis, albeit necessary, would not have been sufficient for the emergence of biochemically functional molecules and, ultimately, life. The mixture of molecules that formed the starting point for the origin of life must have been sorted out by some combination of physical and chemical processes. A. I. Oparin, the Russian biochemist, was the first to propose, in the 1920s, a plausible scenario for how life began. He started from the assumption that organic compounds would

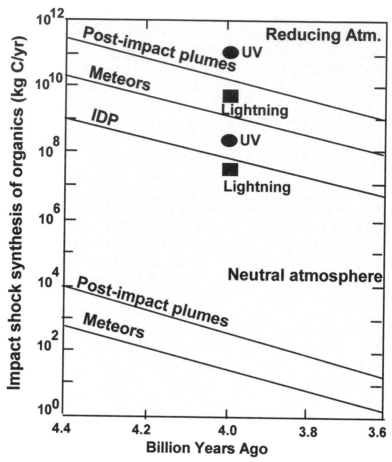

Figure 2.6. Estimated production rates (in kilograms per year) of organic carbon compounds from meteor shockwaves, high-temperature chemical reactions in impact plumes, and in-fall of interplanetary dust particles (IDPs) compared with such production from ultraviolet light and electric discharges. Two sets of data are shown: The lower for a "neutral" (CO_2–N_2–H_2O) atmosphere; the higher for a "reducing" (CH_4–N_2–H_2O) atmosphere. The IDP flux is independent of atmospheric composition. Overall, productivity decreases logarithmically over time as the frequency and size of impactors decreases. (Adapted from Chyba and Sagan 1992.)

have formed spontaneously in a primordial reducing atmosphere, but he also realized that some kind of selection process must have occurred to allow the first stages of prebiotic evolution to begin. This, he suggested, might have been accomplished by the formation of a primitive cell-like entity. Indeed, this remains a viable possibility—the formation of primitive lipids and lipid-based cell-like structures may have played an important role in organizing and simplifying the chaotic mixture of prebiotic compounds initially present. And perhaps minerals—as suggested in the late 1920s by J. B. S. Bernal, a brilliant British biologist who theorized about the origin of life—also played an essential role in the very first stages of such selection on early Earth by selectively adsorbing and/or catalyzing the reactions of certain groups of prebiotic compounds. Today, many researchers would probably agree that a particularly critical event in the origin of life was the appearance of self-replication in some set of information-containing molecules (such as, for example, primitive nucleic acids or proteins). But even after some degree of preselection and simplification had occurred, the spontaneous synthesis of any such highly ordered macromolecule remains a formidable problem.

One important stumbling block for life's origin is the necessity that biological systems be composed of **homochiral** molecules. Many biologically important organic molecules—for example, nucleotides and amino acids—can exist in a multiplicity of structurally different isomeric forms. Some of these structural isomers are sufficiently different that they can be separated by various physical or chemical techniques; however, molecules that differ only in "handedness," or **chirality** (such as D- and L-nucleotides and amino acids), are virtually impossible to separate under normal conditions. Therefore, the earliest prebiotic short polymers of nucleotides (small nucleic acids) or of amino acids (small polypeptides) seem likely to have been composed of random mixtures of dextrorotatory and levorotatory (D and L) isomeric forms. But such mixed, or **heterochiral,** structures are incapable of forming the regular macromolecular structures required for biological activity.

Thus, the origin of living systems appears to have required formation of homochiral macromolecules from a random mix of heteroisomers, chemically a difficult and seemingly unlikely prospect. But there are several possible ways out of this problem. First, it has been suggested, on the basis of physical theory, that under certain prebiotic conditions the D and L isomers of chiral molecules, such as nucleotides and amino acids, need not be formed in equal amounts—although the effect of such discrimination is extremely small and would require amplification by some other mechanism. Second, certain kinds of radiation are known to destroy chiral molecules of a particular handedness at a higher rate than those of the opposite form, a process leading to enrichment in the surviving mix of one of the two isomers. Selective preservation such as this may explain the presence of nonbiologically formed amino acids with slight excesses of D and L isomers in the Murchison meteorite. Third (and most likely), perhaps the molecules themselves "discovered" the answer to the question of chirality.

It has long been suspected that, because of their particular chemical structures, biologically important macromolecules such as DNA or RNA should be capable of directing their own replication so that only one chiral form of their component parts (for nucleotides, the D-isomer rather than its L-enantiomer) would be selected from a mixture of heteroisomers. This has now been demonstrated. In one example, RNA-like molecules were made; although containing the same set of genetic information-storing purines and pyrimidines present in natural RNA, these molecules were built on a different, "unnatural" structural backbone. Sets of short oligomers of these molecules that contained various ratios of D- and L-nucleotide subunits were then compared in terms of the ease with which they carried out a kind of self-replication reaction. The results show that the homochiral sets of molecules replicated faster than the heterochiral sets, suggesting that, in a true competitive process, the all-D- or all-L-containing molecules would ultimately outcompete their slower replicating D/L-mixed competitors. Thus, emergence of order, a prerequisite for the origin of life, can indeed be viewed as a property of purely molecular systems.

It seems inescapable that matter formed in stars now long dead can be traced from the collapse of interstellar clouds, through the evolution of the solar nebula and the assembly and distribution of planetesimals in our Solar System, to the formation of planets and the origin and evolution of life on Earth. Through this long path, the fundamental physical and chemical properties of the biogenic elements that emerged from cosmic evolution have manifested themselves ultimately in the molecular systems engaged in biological evolution on our planet. The collections of meteorites in laboratories and museums offer tangible proof that life is composed of the stuff of stars. Truly, our roots extend far beyond the Sun and deeply into the past. The search for understanding of life's origin in the broadest context affords exciting, indeed virtually unlimited, opportunities for scientific research. Unlike the question of the origin of the Universe, the province mainly of physicists and philosophers, answers to questions bearing on where, when, and how life began can be sought using the tools of virtually all scientific disciplines.

FURTHER READING

Origin of the Elements

Burbidge, E. M., G. R. Burbidge, W. A. Fowler, and F. Hoyle. 1957. Synthesis of the elements in stars. *Reviews of Modern Physics* 29:547–60.

Trimble, V. 1996. The origin and evolution of the chemical elements. In *The origin and evolution of the Universe,* ed. B. Zuckerman and M.A. Malkan, 89–107. Boston: Jones and Bartlett.

Interstellar Dust and Molecules

Gibb, E. L., D. C. B. Whittet, W. A. Schutte, A. C. A. Boogert, J. E. Chiar, P. Ehrenfreund, P. A. Gerakines, J. V. Keane, A. G. G. M. Tielens, E. F. van Dishoeck, and O. Kerkhof. 2000. An inventory of interstellar ices toward the embedded protostar W33A. *Astrophysical Journal* 536:347–56.

Irvine, W. M., and R. F. Knacke. 1989. The chemistry of interstellar gas and grains. In *Origin and evolution of planetary and satellite atmospheres,* ed. S. K. Atreya, J. B. Pollack, and M. S. Matthews, 748–49. Tucson: University of Arizona Press.

Langer, W. D., E. F. van Dishoeck, E. A. Bergin, G. A. Blake, A. G. G. M. Tielens, T. Velusamy, and D. C. B. Whittet. 2000. Chemical evolution of protostellar matter. In *Protostars and planets IV,* ed. V. Mannings, A. Boss, and S. Russell, 29–57. Tucson: University of Arizona Press.

Star and Planet Formation

Andre, P., D. Ward-Thompson, and M. Barsony. 2000. From pre-stellar cores to protostars: The initial conditions of star formation. In *Protostars and planets IV,* ed. V. Mannings, A. Boss, and S. Russell, 59–96. Tucson: University of Arizona Press.

Canup, R. M., and C. B. Agnor. 2000. Accretion of the terrestrial planets and the Earth–Moon system. In *Origin of the Earth and Moon,* ed. R. M. Canup and K. Righter, 113–29. Tucson: University of Arizona Press.

Hollenbach, D. J., H.W. Yorke, and D. Johnstone. 2000. Disk dispersal around young stars. In *Protostars and planets IV,* ed. V. Mannings, A. Boss, and S. Russell, 401–28. Tucson: University of Arizona Press.

Kortenkamp, S. J., E. Kokubo, and S. J. Weidenschilling. 2000. Formation of planetary embryos. In *Origin of the Earth and Moon,* ed. R. M. Canup and K. Righter, 85–100. Tucson: University of Arizona Press.

Mundy, L. G., L. W. Looney, and W.J. Welch. 2000. The structure and evolution of envelopes and disks in young stellar systems. In *Protostars and planets IV,* ed. V. Mannings, A. Boss, and S. Russell, 355–76. Tucson: University of Arizona Press.

Weidenschilling, S. J. 1997. Planetesimals from stardust. In *From stardust to planetesimals,* ed. Y. J. Pendleton and A. G. G. M. Tielens, 281–94. Provo: Brigham Young University Press.

Wuchterl, G., T. Guillot, and J. J. Lissauer. 2000. Giant planet formation. In *Protostars and planets IV*, ed. V. Mannings, A. Boss, and S. Russell, 1081–109. Tucson: University of Arizona Press.

Organic Chemistry of Comets, Interplanetary Dust Particles, Asteroids, and Meteorites

Clemett, S. L., C. R. Maechling, R. N. Zare, P. D. Swan, and R. M. Walker. 1993. Identification of complex aromatic molecules in individual interplanetary dust particles. *Science* 262:721–25.
Cronin, J. R., and S. Chang. 1993. Organic matter in meteorites: Molecular and isotopic analyses of the Murchison meteorite. In *The chemistry of life's origin*, ed. J. M. Greenberg, C. X. Mendoza-Gomez, and V. Pirronello, 209–58. Boston: Kluwer.
Cruikshank, D. P. 1997. Organic matter in the outer Solar System: From the meteorites to the Kuiper belt. In *From stardust to planetesimals*, ed. Y. J. Pendleton and A. G. G. M. Tielens, 315–33. Provo: Brigham Young University Press.
Fomenkova, M. N., S. Chang, and L. M. Mukhin. 1994. Carbonaceous components in comet Halley dust. *Geochimica et Cosmochimica Acta* 58:4503–12.
Irvine, W. M., F. P. Schloerb, J. Crovisier, B. Fegley, and M. J. Mumma. 2000. Comets: A link between interstellar and nebular chemistry. In *Protostars and planets IV*, ed. V. Mannings, A. Boss, and S. Russell, 1159–1200. Tucson: University of Arizona Press.
Mumma, M. J. 1997. Organic volatiles in comets: Their relation to interstellar ices and solar nebula material. In *From stardust to planetesimals*, ed. Y. J. Pendleton and A. G. G. M. Tielens, 369–96. Provo: Brigham Young University Press.
Prinn, R. G., and B. Fegley Jr. 1989. Solar nebula chemistry. In *Origin and evolution of planetary and satellite atmospheres*, ed. S.K. Atreya, J. B. Pollack, and M. S. Matthews, 78–136. Tucson: University of Arizona Press.

Early Geophysical Evolution

Abe, Y., E. Ohtani, T. Okuchi, K. Righter, and M. Drake. 2000. Water in the early Earth. In *Origin of the Earth and Moon*, ed. R. M. Canup and K. Righter, 413–33. Tucson: University of Arizona Press.
Owen, T. C., and A. Bar-Nun. 2000. Volatile contributions from icy planetesimals. In *Origin of the Earth and Moon*, ed. R. M. Canup and K. Righter, 459–71. Tucson: University of Arizona Press.
Shock, E. L., J. P. Amend, and M. Y. Zolotov. 2000. The early Earth vs. the origin of life. In *Origin of the Earth and Moon*, ed. R. M. Canup and K. Righter, 527–43. Tucson: University of Arizona Press.
Sleep, N. H., K. J. Zahnle, J. F. Kasting, and H. J. Morowitz. 1989. Annihilation of ecosystems by large asteroid impacts on the early Earth. *Nature* 342:139–42.

Stevenson, D. J. 1983. The nature of the Earth prior to the oldest known rock record: The Hadean Earth. In *Earth's earliest biosphere, its origin and evolution,* ed. J. W. Schopf, 32–40. Princeton: Princeton University Press.

Walker, J. C. G., C. Klein, M. Schidlowski, J. W. Schopf, D. J. Stevenson, and M. R. Walter. 1983. Environmental evolution of the Archean-Early Proterozoic Earth. In *Earth's earliest biosphere, its origin and evolution,* ed. J. W. Schopf, 260–90. Princeton: Princeton University Press.

Prebiotic Evolution and the Origins of Life

Bolli, M., R. Micura, and A. Eschenmoser. 1997. Pyranosyl-RNA: Chiroselective self-assembly of base sequences by ligative oligomerization of tetranucleotide-2',3'-cyclophosphates (with a commentary concerning the origin of biomolecular homochirality). *Chemical Biology* 4:309–20.

Bonner, W. A. 1998. The quest for chirality. In *Physical origin of homochirality in life,* ed. D. B. Cline, 17–49. Woodbury, N. Y.: American Institute of Physics.

Brack, A., ed. 1998. *Molecular origins of life: Assembling pieces of the puzzle.* Cambridge: Cambridge University Press.

Cairns-Smith, A. G. 1982. *Genetic takeover and the mineral origins of life.* Cambridge: Cambridge University Press.

Chyba, C. F., and C. Sagan. 1992. Endogenous production, exogenous delivery and impact-shock synthesis of organic molecules: An inventory for the origins of life. *Nature* 355:125–32.

Chang, S. 1993. Prebiotic synthesis in planetary environments. In *The chemistry of life's origin,* ed. J. M. Greenberg, C. X. Mendoza-Gomez, and V. Pirronello, 259–99. Boston: Kluwer.

Lahav, N. 1999. *Biogenesis.* Oxford: Oxford University Press.

Shock, E. L. 1990. Geochemical constraints on the origin of organic compounds in hydrothermal systems. *Origins of Life* 20:331–67.

CHAPTER 3

Formation of the Building Blocks of Life

STANLEY L. MILLER AND ANTONIO LAZCANO

INTRODUCTION

Along with his books, notes, letters, and papers, Charles Robert Darwin bequeathed two recipes to succeeding generations. The first, written in his wife's recipe book, describes the way to boil rice:

> Add salt to the water and when boiling hot, stir in the rice. Keep it boiling for twelve minutes by the watch, then pour off the water and set the pot on live coals during ten minutes—the rice is then fit for the table.

His second recipe appears in a letter to his friend Joseph Dalton Hooker (figure 3.1). Sent on February 1, 1871, this letter summarizes Darwin's rarely expressed ideas on the emergence of life and his views on the molecular nature of the basic biological processes:

> It is often said that all the conditions for the first production of a living organism are now present, which could ever have been present. But if (& oh what a big if) we could conceive in some warm little pond with all sorts of ammonia & phosphoric salts,—light, heat, electricity &c present, that a protein compound was chemically formed, ready to undergo still more complex changes, at the present day such matter would be instantly devoured, or absorbed, which would not have been the case before living creatures were formed.

By the time Darwin wrote this letter, DNA had already been discovered, although its central role in genetic processes would not be

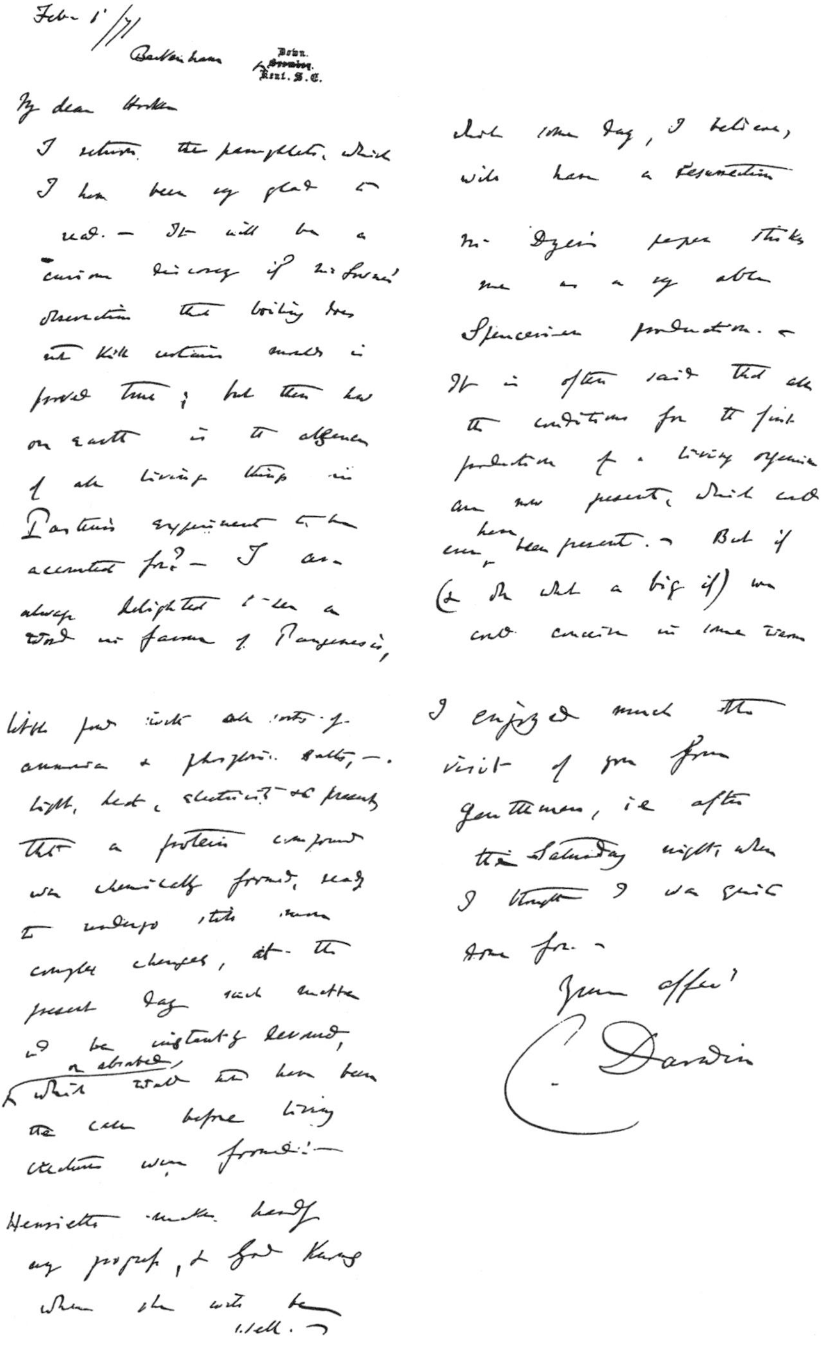

Figure 3.1. Charles Darwin's letter to Joseph Hooker, dated February 1, 1871. In spite of the handwriting, Darwin's ideas on a "warm little pond" are readable. Adapted from Calvin 1969.

TABLE 3.1 DISCOVERY AND CHARACTERIZATION OF
IMPORTANT BIOCHEMICAL MONOMERS

Year	Discoverers	Monomers
1810	W. H. Wollaston	Cystine (urinary calculi)
1819	J. L. Proust	Leucine (fermenting cheese)
1823	M. E. Chevreul	Fatty acids (butyric to stearic)
1869	F. Miescher	DNA (pus cells)
1882	A. Kossel	Guanine (yeast nuclei)
1883	A. Kossel and A. Neumann	Thymine
1886	A. Kossel and A. Neumann	Adenine
1894	A. Kossel and A. Neumann	Cytosine
1900	A. Ascoli	Uracil
1909	P. T. Levene and W. A. Javok	Deoxyribose
1913	W. Küsler	Porphyrin
1906–1936	P. T. Levene et al.	Ribose, ribonucleotides

Based on Leicester (1974) and Letham and Stewart (1977).

deciphered for eighty years. Moreover, the role of proteins in many biological processes had been established, and many of the building blocks of life had been chemically characterized (table 3.1). Of equal significance, by Darwin's day, the chemical gap separating living from nonliving systems had been bridged, at least in principle, by laboratory syntheses of organic molecules. These syntheses challenged the long-entrenched tradition that organic compounds were fundamentally different from inorganic materials. In 1827, for example, Jöns Jacob Berzelius—probably the most influential chemist of his day—wrote that "art cannot combine the elements of inorganic matter in the manner of living nature."

In 1828, however, Berzelius's friend and former student Friedrich Wöhler demonstrated that **urea,** an organic compound fundamental to higher organisms, could be formed in high yield "without the need of an animal kidney" simply by heating ammonium cyanate (Leicester 1974). Wöhler's work represented the first recorded synthesis of an organic compound from purely inorganic starting materials. This benchmark experiment ushered in a new era of chemical research. By 1850, Adolph Strecker had achieved the laboratory synthesis of the amino acid **alanine** from a mixture of acetaldehyde, ammonia, and hydrogen cyanide; this work was followed by the experiments of Alexandr M. Butlerov, who showed that treatment of formaldehyde with a strong alkaline catalyst, such as sodium hydroxide, leads to the formation of **sugars** (table 3.2).

TABLE 3.2 EARLY (19TH-CENTURY) ABIOTIC
SYNTHESES OF BIOCHEMICAL MONOMERS

Starting Materials	Compound(s) Synthesized	Reference
$NH_4NCO \longrightarrow$	urea	Wöhler 1828
$HCH_3CHO + NH_3 + HCN \longrightarrow$	alanine	Strecker 1850
$HCHO \xrightarrow{\quad OH^- \quad}$	sugars	Butlerov 1861

The laboratory synthesis of organic, biochemically relevant compounds was soon extended, offering an increasing array of products and more complicated experimental settings. By the end of the 19th century, electric discharges and diverse gas mixtures had been used to achieve the nonbiological synthesis of sugars and **fatty acids** (Glocker and Lind 1939). Work of this type was carried into the 20th century by several workers, notably Walther Löb and Oskar Baudish, who showed that **amino acids** could be synthesized by exposing wet formamide ($CHONH_2$) either to a silent electric discharge (Löb 1913) or to ultraviolet (UV) light (Baudish 1913). At this time, however, no one recognized that the abiotic synthesis of organic compounds was a prerequisite for the origin of living systems, and no one conceived of these pioneering experiments as laboratory simulations of Darwin's "warm little pond." Instead, working on the general assumption that the earliest forms of life were **autotrophs**—simple plant-like microorganisms that use carbon dioxide (CO_2) rather than organic matter as their carbon source—the scientific community was attempting to understand the autotrophic mechanisms of nitrogen uptake and CO_2 fixation in green plants.

THE HETEROTROPHIC ORIGIN OF LIFE

In the 1860s, Louis Pasteur's swan-necked flask experiments disproved the notion of spontaneous generation. Many believe that the publication of these experiments cut short the scientific discussion of how life began, relegating it to the realm of useless speculation for decades. But the scientific literature of the first part of the 20th century tells a different story. In fact, many major scientists attempted to solve this problem. They proposed a fairly wide spectrum of explanations, ranging from the ideas of Eduard Pflüger on the role of hydrogen cyanide in the origin of life to those of Svante Arrhenius on panspermia. In addition, Leonard Troland hypothesized the existence of a primordial enzyme formed by chance events in the primitive ocean, Alfonso L. Herrera offered a sulfocyanic theory of the origin of cells, and R. B. Harvey suggested a heterotrophic origin of life in a high-temperature environment. And, in a provocative 1926 paper, Hermann J. Muller argued for the abrupt random formation of a single mutable gene endowed with catalytic and autoreplicative properties (for further discussion, see Lazcano 1995).

However diverse, these explanations fell on stony scientific ground. These incomplete speculative schemes, largely devoid of direct supporting evidence, could take no root because they were not subject to fruitful experimental testing. Some of these hypotheses treated life as an emergent feature of nature and thus attempted to understand its origin in an evolutionary context. But in the prevailing view, the first forms of life emerged essentially full-blown as complex **photosynthetic** microbes, endowed from the start with the ability to take in atmospheric CO_2 and combine it with water to synthesize organic compounds.

One proposal, made in the early 1920s by the Russian biochemist Alexandr Ivanovich Oparin, contrasted sharply with the prevalent idea of an autotrophic origin of life. An evolutionary plant biologist trained as a biochemist, Oparin could not reconcile his Darwinian view—gradual slow evolution from the simple to the complex—with the idea that life emerged in a fully endowed form. That is, he could not accept the notion that primordial microorganisms possessed **chlorophyll**, CO_2-capturing enzymes, and all the other biochemical prerequisites for carrying out the complicated process of autotrophic metabolism. Because **heterotrophs** (animals, fungi, and animal-like microbes that obtain energy and grow by breaking down ready-made foodstuffs) are *metabolically* simpler than autotrophs (plants and plant-like microbes that first build foodstuffs, then break them down for energy and growth), Oparin argued that heterotrophs must have evolved before autotrophs. Oparin also realized that anaerobic heterotrophs (microorganisms that use the metabolic process of fermentation to live in the absence of oxygen) are simpler than aerobic (oxygen-requiring) heterotrophs. Thus, he thought that anaerobes must have preceded aerobes. (This idea seemed particularly appealing because the biochemical mechanisms of fermentation are simple; moreover, they are present in all organisms living today—anaerobes and aerobes, both.) Thus, given the simplicity and ubiquity of fermentative metabolism, Oparin (1924) suggested that the first organisms must have been heterotrophic anaerobic bacteria.

Such organisms, he conjectured, did not make their own food; rather, they obtained foodstuffs in the form of organic materials present in the primitive milieu.

A careful reading of Oparin's 1924 pamphlet shows that, at first, he did not assume that the primitive atmosphere was **anoxic** (devoid of molecular oxygen, O_2). Rather, he proposed that **carbides** (carbon–metal compounds) were extruded from the young Earth's interior. Some of these carbides would have reacted with water vapor to form **hydrocarbons** (organic compounds composed solely of atoms of hydrogen and carbon), while others would have oxidized to form more complex oxygen-containing compounds such as aldehydes, alcohols, and ketones (e.g. acetone, CH_3COCH_3). Ammonia (NH_3) would have been present, as well, formed by the hydrolysis of **nitrides** (nitrogen–metal compounds) according to the following reactions:

$$Fe_mC_n + 4mH_2O \rightarrow mFe_3O_4 + C_{3n}H_{8m}$$
$$FeN + 3H_2O \rightarrow Fe(OH)_3 + NH_3$$

Thereafter, the various organic molecules would have reacted among themselves and with ammonia to form "very complicated compounds," ultimately giving rise to proteins and **carbohydrates** (Oparin 1924).

Oparin elaborated and refined his ideas in 1936, publishing them first in Russian and shortly thereafter in English (Oparin 1938). In this new book, he assumed the existence of a highly reducing anoxic environment in which geologically produced iron carbides reacted with steam to form hydrocarbons. These hydrocarbons would then be oxidized (with the oxygen coming from the hot vaporous water) to form aldehydes, alcohols, ketones, and so forth—compounds that would, in turn, react with ammonia to form nitrogen-containing amines, amides, and ammonium salts. The resulting protein-like compounds and other molecules served as constituents of a hot dilute soup within which they aggregated to form simple colloidal globular systems, or **coacervate** droplets. These droplets would evolve into the first heterotrophic microbes. Oparin, like Darwin before him, did not address the formation of nucleic acids (an unsurprising omission, since the role of DNA and RNA in genetic processes was barely suspected at the time).

For Oparin (1938), a "highly reducing" atmosphere corresponded to a mixture of methane (CH_4), ammonia (NH_3), and water vapor (H_2O), with or without additional gaseous hydrogen (H_2)—one of several such reducing mixtures listed in table 3.3. Interestingly, the atmosphere of

TABLE 3.3 TYPES OF PLANETARY ATMOSPHERES

Atmosphere	Composition(s)
Reducing	CH_4, NH_3, N_2, H_2O, H_2
	CO_2, N_2, H_2O, H_2
	CO_2, H_2, H_2O
Neutral	CO_2, N_2, H_2O
Oxidizing	CO_2, N_2, H_2O, O_2

Jupiter contains these very gases, with H_2 in large excess over CH_4. Oparin's proposal of a primordial reducing atmosphere—based, in part, on Vernadsky's idea that the lifeless early Earth must have been anoxic because O_2 is produced by plants—was a brilliant inference derived from a just fledging knowledge of solar atomic abundances and planetary atmospheres. In sum, four benchmark contributions stand out in Oparin's 1938 treatise:

1. The hypothesis that heterotrophs and anaerobic fermentation were primordial

2. The proposal of a reducing atmosphere for the prebiotic synthesis and accumulation of organic compounds

3. The postulated transition from heterotrophy to autotrophy

4. The considerable detail with which these new concepts were addressed

PREBIOTIC SYNTHESIS OF AMINO ACIDS

To buttress his intuition, Oparin needed to demonstrate that organic compounds could form in the absence of living systems. Although he performed no experimental simulations of primordial processes, he recognized that a number of laboratory studies supported his thesis of prebiotic chemical evolution—and hence the heterotrophy of the first life forms. Among these were experiments showing the nonbiological synthesis of sugars (Butlerov's work of 1861), esters, and ethers, as well as the polymerization of glycine ethyl ester and formaldehyde (H_2CO), which Oparin's mentor, A. N. Bakh, investigated. Surprisingly, Oparin's 1938 review mentions neither Strecker's synthesis of alanine (1850) nor Löb's use of electric discharges to synthesize several amino acids (1913). Presumably, Oparin merely overlooked these studies.

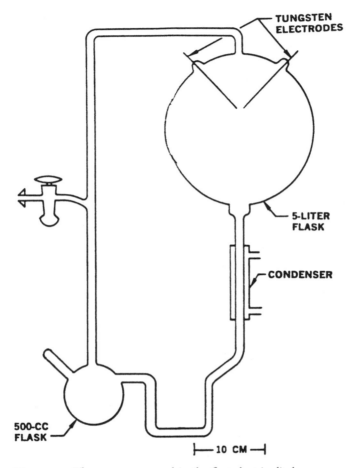

Figure 3.2. The apparatus used in the first electric discharge–
powered synthesis of amino acids and other organic compounds
under conditions designed to simulate those of the primitive
Earth. It was made entirely of glass, except for the tungsten elec-
trodes (Miller 1953).

Firm experimental evidence came first from the laboratory of Harold
Clayton Urey, who was studying the chemical events associated with
origin of the Solar System. Urey (1952) proposed that Earth's primordial atmosphere was hydrogen-rich, and hence highly reducing. In a
laboratory experiment designed to simulate such a setting, the first successful prebiotic amino acid synthesis was achieved using electric discharge as the energy source (figure 3.2) and a strongly reducing model
atmosphere of CH_4, NH_3, H_2O, and H_2 (Miller 1953). Subsequent

TABLE 3.4 ORGANIC COMPOUNDS FORMED BY
SPARKING A GASEOUS MIXTURE OF METHANE,
AMMONIA, WATER VAPOR, AND HYDROGEN

Compound	Yield (mM)	Yield (%)
Glycine	630	2.1
Glycolic acid	560	1.9
Sarcosine	50	0.25
Alanine	340	1.7
Lactic acid	310	1.6
N-Methylalanine	10	0.07
α-Amino-n-butyric acid	50	0.34
α-Aminoisobutyric acid	1	0.007
α-Hydroxybutyric acid	50	0.34
β-Alanine	150	0.76
Succinic acid	40	0.27
Aspartic acid	4	0.024
Glutamic acid	6	0.051
Iminodiacetic acid	55	0.37
Iminoacetic propionic acid	15	0.13
Formic acid	2330	4.0
Acetic acid	150	0.51
Propionic acid	130	0.66
Urea	20	0.034
N-Methyl urea	15	0.051

The percentage yields are based on the amount of carbon present initially as
methane (59 nmoles = 710 mg of carbon).

experiments yielded a score of different amino acids as well as hydroxy
acids, short aliphatic acids, and urea (table 3.4). As expected, the experiments showed that the amino acids were present as **racemic mixtures**
(i.e. each amino acid occurs in two stereoisomeric forms—the form
that occurs in proteins and the nonbiological mirror image of this form).
Surprisingly, however, the experiments did not produce a random mix
of organic molecules; rather, a fairly small number of compounds formed,
and these in substantial quantities. And, with a few exceptions, the
compounds synthesized were of the kinds that make up living systems.

Later investigations examined the precise chemical mechanisms involved in the synthesis of the amino acids and hydroxy acids formed
in the initial spark discharge experiment (Miller 1955). The presence
of large quantities of hydrogen cyanide, aldehydes, and ketones in the
water-filled reaction flask (figure 3.3)—organic compounds clearly

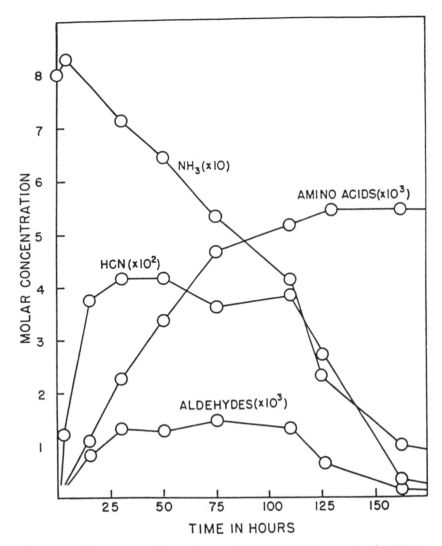

Figure 3.3. The concentrations of ammonia (NH₃), hydrogen cyanide (HCN), and aldehydes (CHO-containing compounds) present in the lowermost U-tube of the apparatus shown in figure 3.2. The concentrations of the amino acids (NH₂- and COOH-containing compounds) present in the lower flask are also shown. These amino acids were produced from the sparking of a gaseous mixture of methane (CH₄), ammonia (NH₃), water vapor (H₂O), and hydrogen (H₂) in the upper flask. The concentrations of NH₃, HCN, and aldehydes decrease over time as they are increasingly converted to amino acids.

derived from the methane, ammonia, and hydrogen in the original reaction mixture—showed that the amino acids were not formed directly in the electric discharge. Rather, their formation was the outcome of a particular set of organic chemical reactions (known to chemists as a "Strecker-like" synthesis) that involved aqueous phase interactions among highly reactive intermediates. The reactions are summarized in scheme 3.1.

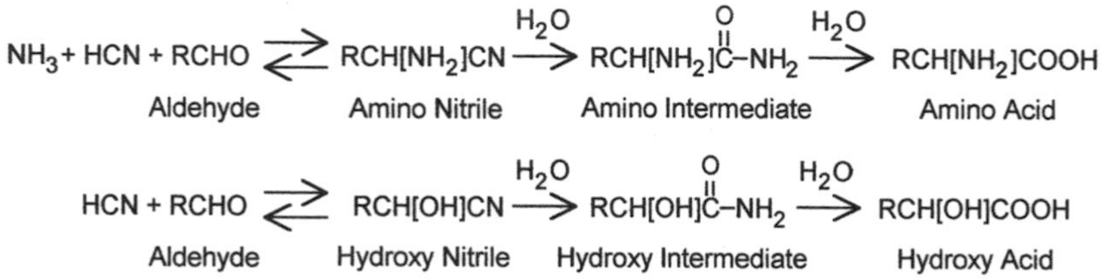

$$NH_3 + HCN + RCHO \rightleftharpoons RCH[NH_2]CN \xrightarrow{H_2O} RCH[NH_2]\overset{\overset{\displaystyle O}{\|}}{C}-NH_2 \xrightarrow{H_2O} RCH[NH_2]COOH$$

| Aldehyde | Amino Nitrile | Amino Intermediate | Amino Acid |

$$HCN + RCHO \rightleftharpoons RCH[OH]CN \xrightarrow{H_2O} RCH[OH]\overset{\overset{\displaystyle O}{\|}}{C}-NH_2 \xrightarrow{H_2O} RCH[OH]COOH$$

| Aldehyde | Hydroxy Nitrile | Hydroxy Intermediate | Hydroxy Acid |

Scheme 3.1.

Detailed studies of the equilibrium and rate constants of these reactions have also been performed (Miller 1955). The results demonstrate that both the amino acids and the hydroxy acids can be synthesized in a simulated primitive ocean even in highly diluted concentrations of hydrogen cyanide (HCN) and aldehyde. The reaction rates depend on temperature and pH as well as on the concentrations of HCN, NH_3, and aldehyde. And they are rapid on a geological time scale: The half-lives for the hydrolysis of the intermediate products in these reactions—amino nitriles and hydroxy nitriles—are less than 1000 years at 0°C (Miller 1998). Moreover, there are no known steps to slow the process. An example of a geologically rapid prebiotic synthesis is that of the amino acids present in the Murchison meteorite (discussed in chapter 2), which apparently took place in less than 10,000 years (Peltzer et al. 1984). These results suggest that if the prebiotic environment of the early Earth was reducing, the synthesis of amino acids was efficient and did not constitute a limiting step in the emergence of living systems.

A **Strecker synthesis** of amino acids on the early Earth would require the presence of the ammonium ion (NH_4^+) in the prebiotic environment. Ammonia is highly soluble in water; thus, if the buffer capacity of the primitive oceans and sediments was sufficient to maintain the pH at 8.1, NH_4^+ would have been available. However, gaseous ammonia is rapidly decomposed by UV light. Thus, the absence of a significant layer of ozone (O_3)—lacking in early Earth history owing to the absence of O_2-producing, photosynthesizing autotrophs—must

TABLE 3.5 PRESENT-DAY VALUES OF FREE ENERGY
SOURCES AVERAGED OVER THE EARTH

Source	Energy per year	
	(cal cm^{-2})	(J cm^{-2})
Total radiation from Sun	260,000	1,090,000
Ultraviolet light		
<3000 Å	3,400	14,000
<2500 Å	563	2,360
<2000 Å	41	170
<1500 Å	1.7	7
Electric discharges	4.0[a]	17
Cosmic rays	0.0015	0.006
Radioactivity (to 1.0 km depth)	0.8	3.0
Volcanoes	0.13	0.5
Shock waves	1.1[b]	4.6

[a]3 cal cm^{-2} of corona discharge plus 1 cal cm^{-2} of lightning per year.
[b]1 cal cm^{-2} of this represents shock waves of lightning bolts (an amount also included as energy from electric discharges).

have imposed an upper limit on the quantity of NH_4^+ resident in the atmosphere. A more realistic atmosphere for the primitive Earth may be a mixture of CH_4 and N_2 with traces of NH_3. Experimental evidence shows that, powered by electric discharges, this mixture of gases is also quite effective in producing amino acids (Miller 1998). Nevertheless, such an atmosphere would be strongly reducing.

A wide variety of direct sources of energy must certainly have been available on the primitive Earth (table 3.5). Among these, solar radiation (rather than atmospheric electricity, simulated by electric discharges) was probably the major source of energy reaching Earth's surface. However, it is very unlikely that any single energy source could account for all of the organic syntheses that occurred. The importance of a given energy source in prebiotic evolution can be calculated by multiplying the estimated amount of available energy by its established efficiency for organic synthesis. Given our current ignorance of the prebiotic environment (we have no direct geologic evidence for several hundred million years of Earth's history), it is impossible to make absolute assessments of the relative significance of each energy source. For instance, neither **pyrolysis** (a breakdown of compounds due to intense heating) at 800 to 1200°C of a mixture of CH_4 and NH_3, nor the action of UV light on a strongly reducing atmosphere, yields large amounts of amino acids. But pyrolysis of a mixture of methane, ethane

(C_2H_6), and other hydrocarbons gives good yields of phenylacetylene (C_8H_6), whose hydration produces phenylacetaldehyde. And in the prebiotic ocean, phenylacetaldehyde could have participated in a Strecker synthesis as a precursor to the amino acids phenylalanine and tyrosine.

Solar radiation may have been the major energy source reaching Earth's surface, but electric discharges were evidently the most important source of HCN. Hydrogen cyanide is universally recognized as a crucial intermediate in many prebiotic organic syntheses, including the formation of amino nitriles by Strecker-type syntheses. In addition, HCN polymers also serve as sources of amino acids. Ferris et al. (1978) established that, in addition to being a source of urea, guanidine, and oxalic acid, HCN polymers can hydrolyze to form glycine, alanine, aspartic acid, and α-aminoisobutyric acid; however, the reported yields of these amino acids are not high, the largest being that of glycine, at approximately 1%.

Amino Acid Synthesis in
Mildly Reducing and Nonreducing Atmospheres

We are fairly certain that free oxygen was absent from Earth's primordial environment. But the composition of the primitive atmosphere remains elusive. Suggested atmospheres span the range from strongly reducing ($CH_4 + NH_3 + H_2O$, $CO + N_2 + H_2O$, or $CO_2 + N_2 + H_2$) to chemically neutral ($CO_2 + N_2 + H_2O$). For the most part, atmospheric scientists tend to favor neutral (nonreducing) compositions whereas specialists in prebiotic chemistry favor a more reducing makeup, a composition that renders abiotic synthesis of amino acids especially efficient. (In contrast, the carbon source—CH_4, CO, or CO_2—would not affect the prebiotic synthesis of purines and sugars as long as sufficient H_2 remained available.)

Unfortunately, gas mixtures containing CO and CO_2 have received substantially less experimental attention than have those containing CH_4. Spark discharge experiments using CH_4, CO, or CO_2 as a carbon source together with various amounts of H_2 have shown that the mixtures containing methane produce the highest yields of amino acids (figure 3.4), but experiments using CO or CO_2 are almost as productive if the H_2/C ratio in the gas mixture is high. Without an excess of hydrogen, however, the amino acid yields are very low, especially when CO_2 is the sole carbon source. The suite of amino acids produced in CH_4-based experiments (listed in table 3.4) is similar to that first reported by Miller (1953). But in experiments using CO or CO_2, the

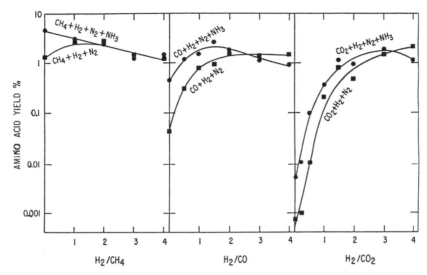

Figure 3.4. Yields of amino acids (as a percentage of initial carbon) produced in three sets of electric discharge experiments carried out in the apparatus shown in figure 3.2. The experiments used methane (CH_4), carbon monoxide (CO), or carbon dioxide (CO_2) as the carbon source, together with hydrogen (H_2), nitrogen (N_2), and, in some cases, ammonia (NH_3). Regardless of carbon source, all the experiments produced similar yields of amino acids. The experiments were conducted at room temperature over 2-day periods of continuous spark discharge; partial pressures of N_2, CO, or CO_2 were 100 torr; the lower flask contained 100 ml of water but no NH_3.

simplest amino acid, glycine, predominates; little else is produced except for small amounts of alanine (Miller 1998).

Such results imply that CH_4 is the best carbon source for abiotic syntheses, especially of amino acids. But what if CO or CO_2 were predominant in the early atmosphere? Glycine is virtually the only amino acid produced in spark discharge experiments using CO or CO_2; but as the primitive ocean matured, reactions between glycine ($CH_2[NH_2]COOH$), formaldehyde (H_2CO), and hydrogen cyanide (HCN) would have generated other amino acids such as alanine, aspartic acid, and serine. For example, at pH > 9, synthesis of serine is favored at H_2CO concentrations greater than 10^{-3} M, as shown in scheme 3.2.

$$CH_2[NH_2]COOH \xrightarrow{OH-} CH[NH_2]COOH \xrightarrow{CH_2O} CH_2[OH]CH[NH_2]COOH$$

Glycine   Serine

Scheme 3.2.

It is possible that the simple mixtures of organic chemicals made using CO or CO_2 as a carbon source lacked the diversity required for prebiotic evolution. However, we do not know which of the 20 biologically common amino acids were actually required for the emergence of life. We can say that although CO and CO_2 are less favorable than CH_4 for amino acid prebiotic synthesis, the amino acids produced from CO and CO_2 may have been adequate. Interestingly, under some conditions, the spark discharge yields of amino acids, HCN, and aldehydes are about the same whether the gas mixtures used CH_4, CO (at $H_2/CO > 1$), or CO_2 (at $H_2/CO_2 > 2$) as their sole source of carbon. However, H_2 gravitationally escapes from Earth's atmosphere, so it is not clear how the prebiotic atmosphere could have maintained the high hydrogen-to-carbon ratios demanded by the last two reaction mixtures.

PREBIOTIC SYNTHESIS OF NUCLEIC ACID BASES

Joseph Louis Proust was a French chemist whose belief in the "law of constant proportions" kept him busy purifying and analyzing various chemical compounds. In 1807, while teaching in Madrid, he undertook a study of aqueous solutions of HCN. He found that if the solutions were basic, having a pH > 7.0, a complex polymer was produced, together with several other uncharacterized compounds. In retrospect, it seems quite possible that adenine, one of the components of nucleic acids, was one of Proust's uncharacterized side-products. Proust's work notwithstanding, it was 1960 before the world understood that the components of nucleic acids could be synthesized abiotically. At that time, John Oró was studying the synthesis of amino acids from aqueous solutions of HCN and NH_3. In a landmark contribution, he reported the abiotic formation of adenine—one of the two purines in the nucleic acids of all living systems (Oró 1960).

Oró's synthesis of adenine is indeed remarkable. If concentrated solutions of ammonium cyanide are refluxed for a few days, adenine is obtained in as much as 0.5% yield, along with the side-products 4-aminoimidazole-5-carboxamide and a cyanide polymer (Oró 1960; Oró and Kimball 1961, 1962). The probable mechanism of this synthesis is shown in scheme 3.3.

In this sequence, the limiting reaction is the penultimate step, in which the HCN tetramer (diaminomaleonitrile) combines with formamidine to form aminoimidazole carbonitrile. However, as demonstrated by Ferris and Orgel (1966), this step can be bypassed by a two-photon

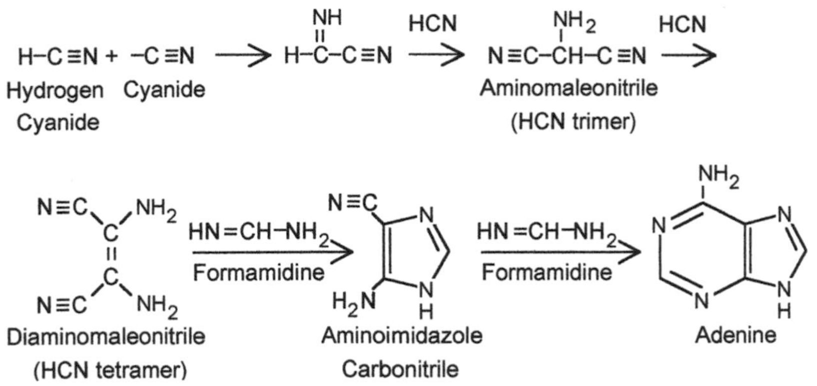

Scheme 3.3.

photochemical rearrangement of diaminomaleonitrile (scheme 3.4), which proceeds readily in sunlight to give high yields of aminoimidazole carbonitrile (AICN):

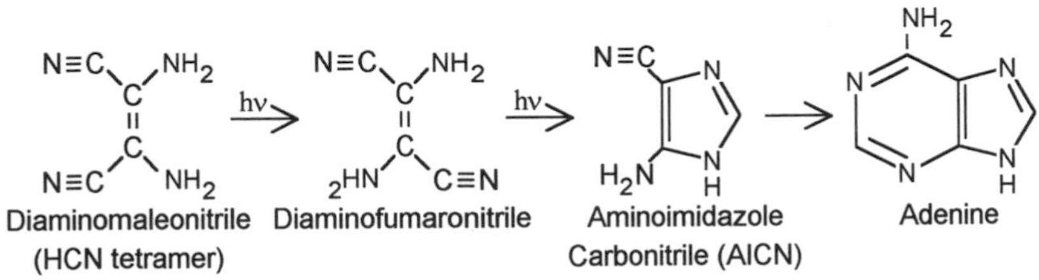

Scheme 3.4.

On the primitive Earth, formation of the tetramer (diaminomaleonitrile) may have been accelerated in a **eutectic** solution (one at its lowest possible melting point) of HCN and H_2O, which could reasonably have existed in polar regions. High yields of the HCN tetramer have been reported from experiments in which dilute cyanide solutions were cooled for a few months to temperatures between -10 and $-30°C$ (Sanchez, Ferris, and Orgel 1966a). Production of adenine by the polymerization of HCN is also speeded by the presence of formaldehyde and other aldehydes; these, too, were probably available in the prebiotic environment (Voet and Schwartz 1983).

The prebiotic synthesis of guanine, the other purine of biological nucleic acids, was first studied under conditions that required unrealistically high concentrations of several precursors, including ammonia (Sanchez, Ferris, and Orgel 1967). On the basis of subsequent findings, it was then proposed that, together with guanine, other purines (including hypoxanthine, xanthine, and diaminopurine) may have been produced in the primitive environment by variations of Oró's adenine synthesis—the result of reactions of aminoimidazole carbonitrile and aminoimidazole carboxamide, as shown in scheme 3.5 (Sanchez, Ferris, and Orgel 1968).

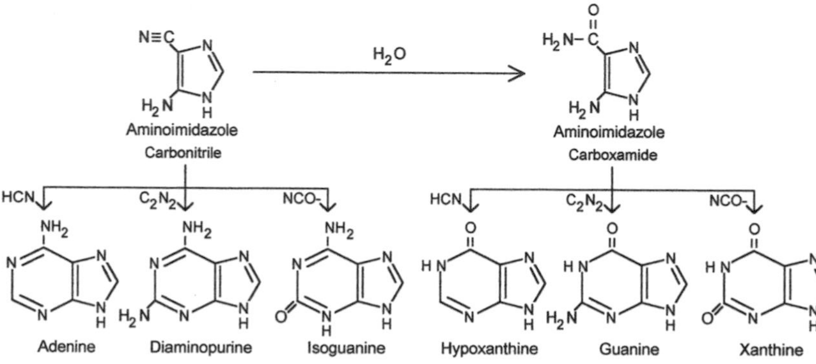

Scheme 3.5.

A re-examination of the polymerization of concentrated solutions of NH_4CN (ammonium cyanide) has shown that, in addition to adenine, guanine is produced both at -80 and at $-20°C$ (Levy, Miller, and Oró 1999). It is likely that most of the guanine obtained from the polymerization of NH_4CN is a product of diaminopurine, which reacts readily with water, undergoing hydrolytic deamination to give guanine and isoguanine. The yields of guanine from this simple one-pot reaction synthesis of the various purines are 10 to 40 times less than those of adenine. A possible reaction sequence for this synthesis is shown in figure 3.5. The mechanisms depicted are very likely oversimplified. Although adenine can be considered a pentamer of HCN, in dilute aqueous solutions, its synthesis involves the formation and rearrangement of other precursors such as 2-cyanoadenine and 8-cyanoadenine (Voet and Schwartz 1983).

The abiotic synthesis of cytosine, one of the principal pyrimidine nucleic acid bases, has also been described (Sanchez et al. 1966b; Ferris et al. 1968). It is made in aqueous phase from cyanoacetylene (HC_3N) and cyanate (NCO^-). Cyanoacetylene is produced abundantly by the action of a spark discharge on a mixture of methane and nitrogen, and cyanate can come from cyanogen (CN) or from the decomposition of urea ($CO[NH_2]_2$). However, the high concentrations of NCO^- (>0.1 M) required by this reaction are unrealistic, because cyanate would rapidly hydrolyze to CO_2 and NH_3 in the primitive ocean.

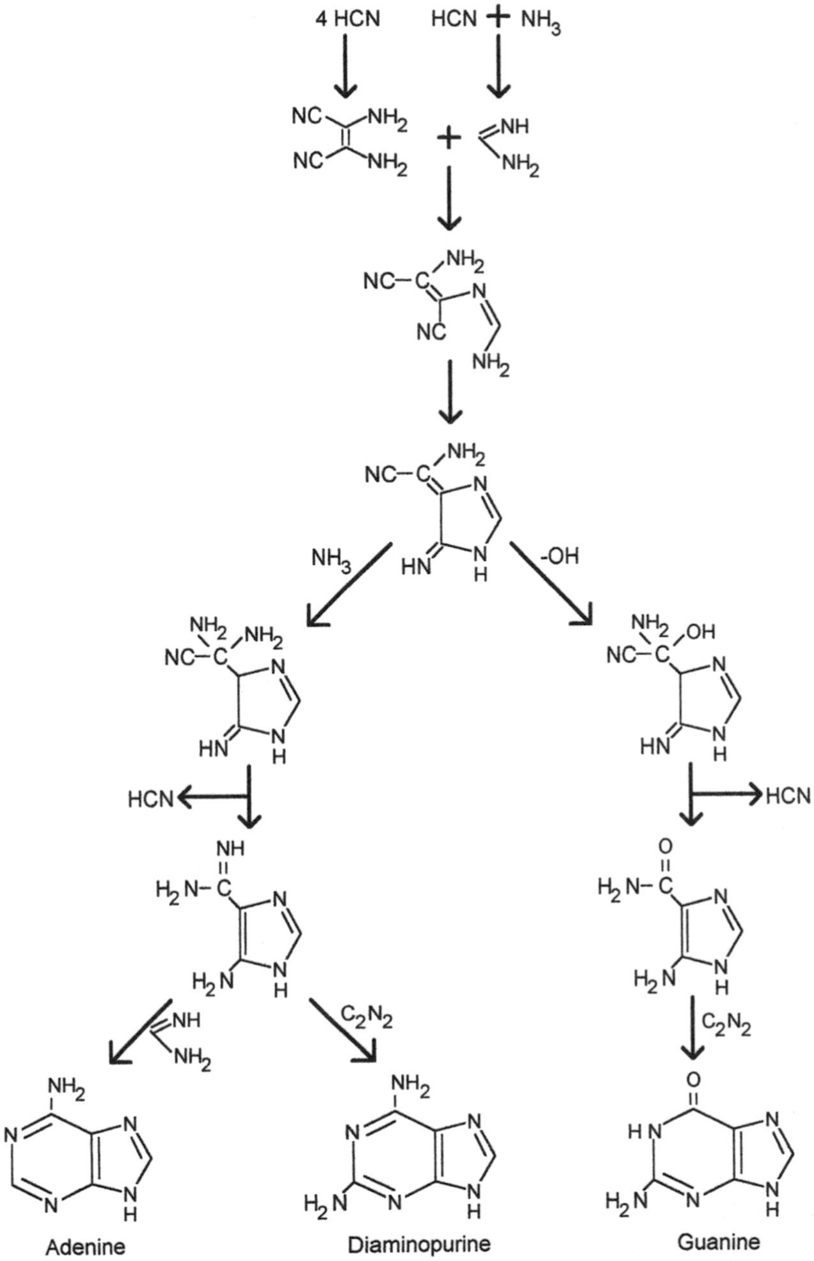

Figure 3.5. Possible mechanisms for prebiotic synthesis of the purines adenine, diaminopurine, and guanine from a mixture of hydrogen cyanide (HCN) and ammonia (NH₃). Guanine would also be formed from the hydrolysis of diaminopurine (not shown).

More plausible mechanisms for the abiotic synthesis of pyrimidines have also been suggested. Orotic acid, a biosynthetic precursor of the pyrimidine uracil, has been identified (albeit in low yields) among the hydrolytic products of hydrogen cyanide polymers (Ferris et al. 1978). The reaction of cyanoacetaldehyde (produced in high yields from the hydrolysis of cyanoacetylene) with urea produces no detectable levels of the pyrimidine cytosine (Ferris et al. 1974). But when the same nonvolatile compounds are concentrated in an evaporating pond (simulating primitive evaporating lagoons on the early Earth), notably high amounts of cytosine (>50%) are formed (Robertson and Miller 1995). The results suggest a facile mechanism for the accumulation of the pyrimidines cytosine and uracil in the prebiotic environment, as outlined in scheme 3.6.

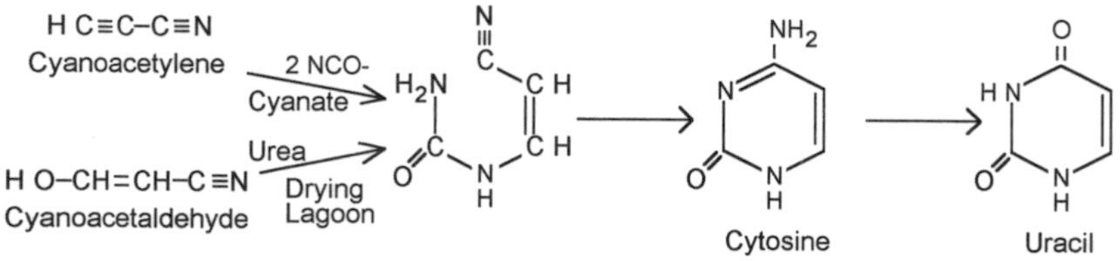

Scheme 3.6.

A related synthesis under evaporative conditions has been used to react cyanoacetaldehyde with guanidine to produce **diaminopyrimidine** in very high yields (Ferris et al. 1974; Robertson et al. 1996). Upon hydrolysis, this compound produces uracil together with very low yields of cytosine. Although it is unlikely that high amounts of diaminopyrimidine were present in the prebiotic environment, a wide variety of other modified nucleic acid bases may have existed. The list includes isoguanine, a hydrolytic product of diaminopurine (Levy et al. 1999), and various modified purines. Such compounds may be the results of side reactions when adenine and guanine react with a number of different amines under the concentrated conditions of a drying pond (Levy and Miller 1999). Other modified purines are shown in figure 3.6.

Chemically modified pyrimidines, formed efficiently under plausible prebiotic conditions, may also have been present in the early environment (Robertson and Miller 1995b). These include such compounds as dihydrouridine (formed from NCO⁻ and β-alanine), diaminopyrimidine, thiocytosine (Robertson et al. 1996), and 5-substituted uracils, whose functional side groups (figure 3.7) may have played an important role in the early evolution of catalysis before the origin of proteins.

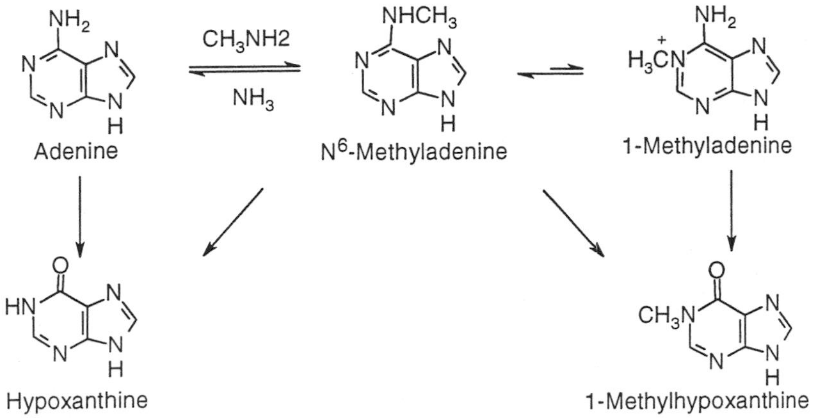

Figure 3.6. Prebiotic synthesis from adenine of various modified purines (N^6-methyladenine, 1-methyladenine, hypoxanthine, and 1-methylhypoxanthine).

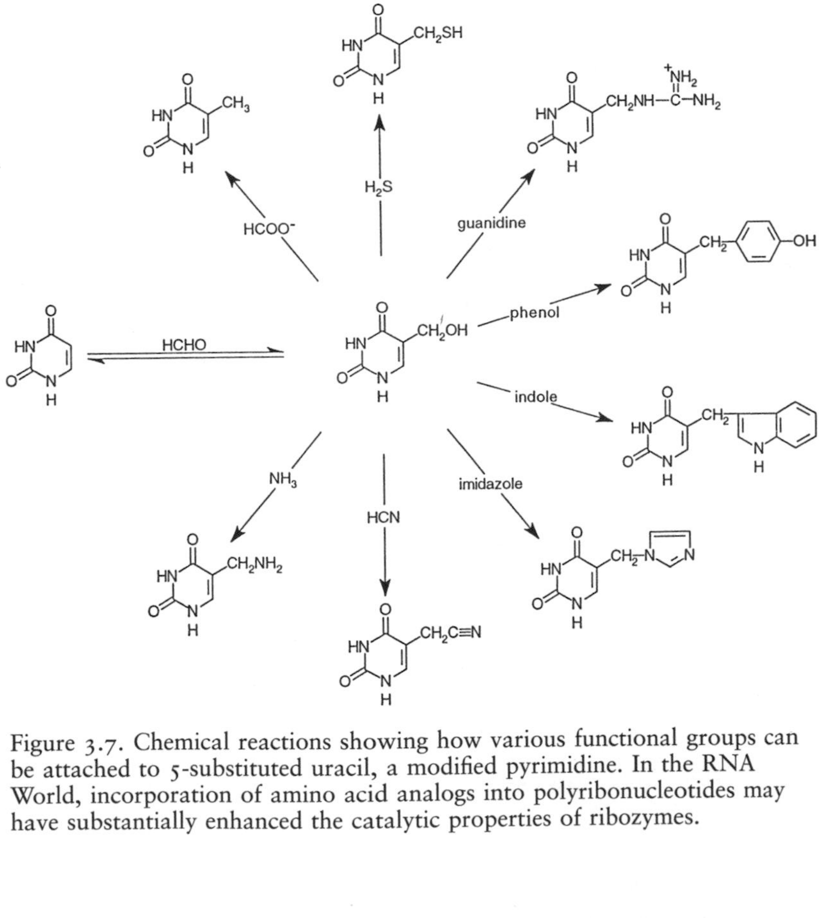

Figure 3.7. Chemical reactions showing how various functional groups can be attached to 5-substituted uracil, a modified pyrimidine. In the RNA World, incorporation of amino acid analogs into polyribonucleotides may have substantially enhanced the catalytic properties of ribozymes.

PREBIOTIC SYNTHESIS OF SUGARS

Ribose may have been one of the sugars employed by the earliest living systems. Its importance was recognized with the discovery of the **RNA World,** a stage of early biological evolution when living systems used RNA both as a catalyst and as an informational macromolecule. And, together with a great many other sugars produced by the condensation of formaldehyde under alkaline conditions (Butlerov 1861), ribose can be synthesized in the laboratory under conditions relevant to the prebiotic Earth.

However, the **Butlerov synthesis** of sugars, also known as the **formose reaction,** is very complex. It depends on the presence of a suitable inorganic catalyst, most commonly calcium hydroxide (Ca[OH]$_2$) or calcium carbonate (CaCO$_3$). In the absence of such inorganic catalysts, little or no sugar is produced. At 100°C, clay minerals such as kaolin also catalyze the formation of small yields of simple sugars, including ribose, from dilute (0.01 M) solutions of formaldehyde (Gabel and Ponnamperuma 1967; Reid and Orgel 1967).

The Butlerov synthesis is **autocatalytic**—that is, catalyzed by its own products. It proceeds in a series of steps from formaldehyde through glycoaldehyde, glyceraldehyde, and dihydroxyacetone (four-carbon sugars), to pentoses (five-carbon sugars), to hexoses (six-carbon sugars) such as glucose and fructose. These six-carbon, simple sugars are important constituents of biological carbohydrates. The detailed reaction sequence is not yet understood, but may proceed as shown in scheme 3.7.

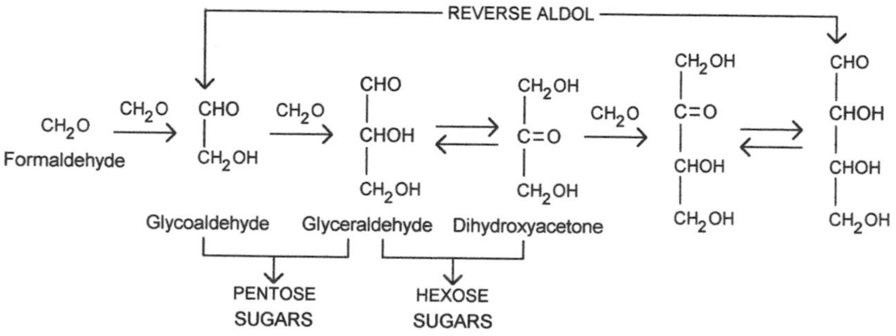

Scheme 3.7.

Two major problems attend the assumption that a Butlerov-style reaction played a pivotal role in producing sugars on the primitive Earth. First, the Butlerov synthesis is *too* efficient: It generates a great variety of sugars, both straight-chain and branched (figure 3.8). Indeed,

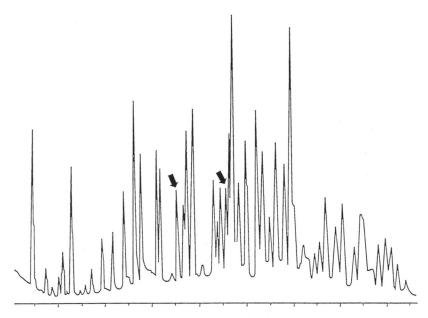

Figure 3.8. Gas chromatogram in which each of the peaks reflects the presence of a sugar formed by a Butlerov synthesis (a formose reaction). The arrows point to peaks produced by the two forms of ribose formed in the reaction: D-ribose, the form present in RNA, and L-ribose, its nonbiological mirror image. Adapted from Decker et al. (1982).

more than 40 different sugars, including small amounts of ribose, have been identified as products in a single reaction mixture (Decker et al. 1982). Second, the ribose formed is a racemic mixture, consisting of D-**ribose** (the configuration found in the biological nucleic acids RNA and DNA) and its mirror image L-**ribose** (a form not found in biological systems). All sugars have fairly similar chemical properties; thus, it is difficult to envision simple physicochemical mechanisms that could (1) preferentially concentrate ribose from a complex mixture or (2) enhance the yield of the D-ribose relative to that of its biologically inactive mirror image.

Albert Eschenmoser and his colleagues have proposed a promising solution of these problems (Muller et al. 1990). They recognized that in biological systems, sugar biosynthesis generates sugar-phosphate compounds rather than free, unbound sugars. Under slightly basic conditions, condensation of glycoaldehyde phosphate in the presence of formaldehyde leads to considerable selectivity in the synthesis of ribose-2,4-diphosphate, a compound that exists either as a six-membered ring or a

TABLE 3.6 RATE OF DECOMPOSITION OF VARIOUS
SUGARS (100°C, 0.05 M PHOSPHATE)

Sugar	Time$_{1/2}$ (min)
Ribose	73
Ribose (0.05 M HCO_3^-)	140
2-Deoxyribose	225
Ribose-5-phosphate	7
Ribose-2,4-diphosphate	31

pyranose diphosphorylated sugar (Muller et al. 1990). In the presence of minerals, this reaction also takes place at neutral pH and low reactant concentrations (Pitsch et al. 1995). Judging from the properties of pyranosyl-RNA (discussed in chapter 4), a 2′,4′-linked nucleic acid analog whose backbone includes the six-membered pyranose form of ribose-2,4-diphosphate seems a particularly attractive prebiotic pathway.

The inherent instability of ribose poses yet another problem with respect to its prebiotic availability. Under neutral conditions (pH 7), the half-life for the decomposition of ribose is 73 minutes at 100°C and only 44 years at 0°C (Larralde, Robertson, and Miller 1995). As summarized in table 3.6, other pentose and hexose sugars are similarly unstable, as is ribose-2,4-diphosphate. Although many ways have been suggested to stabilize sugars, attaching the sugar to a purine or pyrimidine—that is, linking the sugar to a **nucleoside**—is the most biologically interesting. But the synthesis of such sugar–base nucleosides is notoriously difficult to achieve under truly prebiotic conditions. Thus, ribose-containing nucleosides are unlikely to have been components of the earliest prebiotic informational macromolecules (Shapiro 1988). As discussed in chapter 4, various workers have proposed possible substitutes for ribose as simpler precursors for the informational macromolecules known collectively as **pre-RNAs**.

HYDROTHERMAL VENTS AND THE ORIGIN OF LIFE

The discovery of hydrothermal vents at submarine ridge crests was a major development in oceanography, engendering an appreciation of their effect on the ionic balance of the world's oceans (Corliss et al. 1979). Large amounts of ocean water now pass through such vents—

in fact, a volume equivalent to the world's entire ocean system circulates through them every ten or so million years (Edmond et al. 1982). Moreover, such vents have a long history. Hydrothermal vents and the accompanying processes of hydrothermal oceanic circulation almost certainly date from Earth's beginnings, when the flow was even greater because heat flow from the planet's interior was especially high (see chapter 2).

Soon after the discovery of deep-sea vents, Corliss and his colleagues (1981) proposed a detailed hypothesis for the origin of life in this abyssal hydrothermal setting. They suggested that amino acids and other organic compounds were produced by chemical reactions occurring during passage of waters through the vent system, where the temperature gradient ranged from 350°C (in the vent waters) to ~2°C (in the surrounding ocean). Such conditions favor polymerization and permit the resulting polymers to self-organize, leading to the development of living systems.

Given the geological plausibility of a hot early Earth, submarine hydrothermal springs appear to be an ideal setting for the origin of life. We know that more than a hundred such vents exist today along active tectonic oceanic ridge crests. And, in at least some of these, huge amounts of catalytic clays and minerals interact with an aqueous reducing environment rich in H_2, H_2S, CO, CO_2 (and possibly HCN, CH_4, and NH_3)—simple compounds known to react under prebiotic conditions to form amino acids, purines, pyrimidines, and other biochemicals. Moreover, until recently, the possibility that life originated in high-temperature environments (Pace 1991), including those found at deep-sea vents, appeared to be supported by molecular phylogenetic trees, whose earliest diverging branches are occupied by anaerobic microbial **hyperthermophiles**—microorganisms that grow optimally at and above 90°C (Stetter 1994). However, current studies suggest that hyperthermophily is actually a secondary adaptation that evolved early in Earth's history but well after the origin of life. Galtier, Tourasse, and Gouy (1999), using a newly developed method for the statistical analysis of the G + C (guanine plus cytosine) content of the DNA genes that code for **rRNA** (ribosomal RNA)—for biological systems, a reliable indicator of environmental temperature—have concluded that the **Last Common Ancestor** (LCA) of extant life (the evolutionary rootstock of all present-day organisms) was not a hyperthermophile. Rather, the LCA was part of an ancestral population of **mesophiles** (organisms inhabiting settings with moderate temperatures).

That life may have originated under thermophilic conditions is not a new idea. It was first proposed by Harvey (1924), who argued that the earliest life forms were heterotrophic **thermophiles** that originated in hot springs like those at Yellowstone National Park (Wyoming, U. S.). As Harvey emphasized, chemical reactions would have proceeded faster in a high-temperature setting, and primitive enzymes would have been less efficient than their modern counterparts. However, high temperatures are also destructive to organic compounds. Among other deleterious effects, the consequent loss of biochemical structure by decomposition diminishes the stability of genetic material such as RNA. As noted earlier, ribose and other sugars are notably thermolabile (Larralde et al. 1995). In a high-temperature environment, could ribose survive long enough to participate in the prebiotic synthesis of the macromolecules required by the RNA World?

The stability of ribose and other sugars is a serious problem (table 3.6), but that of pyrimidines, purines, and some kinds of amino acids present difficulties as well. At 100°C, the half-life for deamination of the pyrimidine cytosine is only 21 days, and for the purine adenine, 204 days (Shapiro 1995). Amino acids present a mixed case. Some, like alanine, are thermally quite stable; but others, like serine, are not. At 100°C, alanine has a decarboxylation half-life of ~19,000 years whereas serine's decarboxylation half-life is 320 days (Vallentyne 1964). Taken together, these observations demonstrate that neither the growth of organisms nor the origin of life is at all likely at 250 or 350°C (White 1984; Miller and Lazcano 1995).

In summary, it is difficult to accept that organic compounds were synthesized at 350°C in submarine vents. Rather, we have abundant data indicating that such conditions favor decomposition of many compounds in time spans ranging from seconds to hours. Thus, the suggestion that life originated in such a setting (Corliss et al. 1981) is highly improbable. This is not to imply that deep-sea hydrothermal springs were negligible factors in life's beginnings. If the mineral assemblages at such vents were sufficiently reducing, the vents could have been a source of CH_4, which, diffused into the atmosphere, may have served as a carbon source for organic syntheses. And such hydrothermal vents would also have played a role in regulating the ionic composition of the ocean, as they do today. Finally, because submarine vent systems could destroy organic compounds, their presence would have fixed an upper limit on the concentration of organics in the primitive ocean.

EXTRATERRESTRIAL ORGANIC
COMPOUNDS AND PREBIOTIC EVOLUTION

The existence of extraterrestrial organic compounds has been recognized since 1834, when Berzelius analyzed constituents of the Aläis carbonaceous meteorite (now classed as a C1 carbonaceous chondrite), a conclusion confirmed a few years later by Wöhler's studies of the Kaba meteorite (a C2 carbonaceous chondrite). Today, the presence of a complex array of extraterrestrial organic compounds in meteorites, comets, interplanetary dust, and interstellar clouds is firmly established. Some workers have proposed that extraterrestrial material represents a promising source of the prebiotic organic compounds necessary for the origin of life (Oró 1961; Anders 1989; Chyba 1990; Chyba and Sagan 1992). Positing an extraterrestrial origin for the organic components of Earth's prebiotic soup has a singular appeal: It accounts for the low yields and limited variety of organic compounds synthesized in the CO_2-rich models of primitive atmospheres favored by atmospheric scientists (Kasting 1993).

But the source of organic compounds in the primordial broth is not the question. Instead, we must ask *what percentage* came from each of several potential sources. Numerous types of interstellar molecules have been detected (including formaldehyde, hydrogen cyanide, acetaldehyde, cyanoacetylene, and other prebiotic compounds), and their total amount in any particular interstellar cloud is high. However, their concentrations are very low (a few molecules per cubic meter), and there are no plausible mechanisms to concentrate and deliver them to Earth. Thus, interstellar molecules could have played, at most, a minor role in the origin of life.

The major sources of exogenous organic compounds appear to be comets and interplanetary dust particles, asteroids and carbonaceous meteorites being minor contributors. Asteroids would have impacted the early Earth frequently, especially during the planet's earliest several hundred million years; but the amount of organic material brought in would have been rather small, even if such asteroids were similar in composition to the organic-rich Murchison meteorite. Murchison contains a low percentage of organic carbon, mostly in the form of an insoluble carbonaceous polymer, as well as about 100 ppm of amino acids, a minuscule amount ($\sim$0.10 g/kg of meteorite, or $\sim$2.0 $\times$ 10^{-2} moles of amino acids, assuming a void volume for the meteorite of

10% and a density of approximately 2.0). Consider, for example, the asteroid, some 10 km across, whose crash was instrumental in the extinction of the dinosaurs at the end of the Cretaceous Period (65 million years ago). The amino acids carried in on this huge body were present in such minute amounts that they were detectable in the rock record only by the most sensitive of modern analytical techniques (Zhao and Bada 1989). The contribution of organic compounds carried in on carbon-rich meteorites would have been equally minor. And their contribution to the primordial soup could never be more than minor, even if a significant proportion of meteoritic amino acids, pyrimidines, and purines survived the high temperatures associated with frictional heating during atmospheric entry (Glavin and Bada 1999).

Among the various sources of prebiotic exogenous organic compounds suggested, comets are the most promising (Oró and Lazcano 1997), a proposal first made by John Oró (1961). Cometary nuclei contain about 80% H_2O, 1% HCN, and 1% H_2CO, as well as significant amounts of CO_2 and CO. Thus, if we assume that a typical cometary nucleus has a density of 1 g cm^{-3}, a comet 1 km across would contain $\sim 2 \times 10^{11}$ moles of HCN, an amount equivalent to about 40 nmoles cm^{-2} distributed over Earth's surface. This is a huge amount, comparable to the yearly production of HCN by electric discharge in a CH_4-rich reducing atmosphere. Such an abundance would certainly have played a major role in organic syntheses if the early Earth lacked a reducing atmosphere. (This calculation assumes a complete survival of HCN on impact. But we have little understanding of what happens during the impact of such an object, much of which would be heated to temperatures above 300°C. This is high enough to decompose HCN and any other organic compounds brought in. However, such decomposition could yield highly reactive chemical species that might serve as precursors in the abiotic synthesis of prebiotic organic monomers.)

Interplanetary dust particles also carry in organics. Today, the annual in-fall rate of such particles is $\sim 40 \times 10^6$ kg (Love and Brownlee 1993). On primitive Earth, when the planet was forming, this amount may have been 100 to 1000 times greater. Unfortunately, the organic composition of such cosmic dust is poorly known. To date, the only kinds of organic molecules identified are a few types of polycyclic aromatic hydrocarbons (Gibson 1992; Clemett et al. 1993). But much of the dust may consist of **tholins**—organic polymers of a type produced by electric discharge, ionizing radiation, and ultraviolet light in laboratory experiments. Although tholins are generally quite refractive, they release some amino acids (a low percentage) upon acid hydrolysis. More promisingly, tholins may be a source of prebiotic chemical precursor molecules such as hydrogen cyanide, cyanoacetylene, and various aldehydes. On entry into Earth's primitive atmosphere, the tholin-carrying dust particles would be heated, pyrolyzing the organic matter. Then the resulting prebiotic compounds, including HCN, could have participated in a sequence of prebiotic reactions (Mukhin et al. 1989; Chyba et al. 1990).

The true source of organics on the primitive Earth remains an open question. But even if we could prove that comets and meteorites delivered them, we could not confirm the concept of panspermia. Instead, we could say with certainty that the primitive terrestrial environment was partly shaped by the same intense bombardment that affected the Moon, Mars, and other bodies of the Solar System.

AN AUTOTROPHIC ORIGIN OF LIFE?

Alternative theories explaining the origin of life coexist today. Some argue for an RNA World (Gilbert 1986) or a **Thioester World** (De Duve 1991), others maintain that life originated at submarine hydrothermal vents (Corliss et al. 1981), and still others hypothesize an extraterrestrial origin for the requisite organic compounds (Oró 1961). Underlying all these notions is the shared assumption that nonbiological organic synthesis was a necessary precursor to the emergence of living systems. Moreover, each is consistent with a heterotrophic origin of life (table 3.7). But a competing idea—an autotrophic origin of life—has proponents, as well. According to this theory, the earliest forms of life were plant-like autotrophs (they used CO_2 as a source of carbon). These

TABLE 3.7 CURRENT HYPOTHESES ON THE ORIGIN
OF LIFE (EXCLUDING PANSPERMIA)

1. Abiotic Synthesis and Heterotrophic Origin

Oparin (1936)	Primitive soup and primordial fermentation
Corliss et al. (1981)	Submarine hot spring thermophilic heterotroph
Gilbert (1986)	RNA World
De Duve (1991)	Thioester World
Kauffman (1993)	Self-organization and complexity theory

2. Primordial CO_2 Fixation and Autotrophic Origin

Wächterhäuser (1988)	Pyrite-based chemolithotrophic metabolic networks

emergent chemical systems synthesized their own organic compounds instead of incorporating ready-made environmental organics. The most articulate proponent of the autotrophic hypothesis is Günter Wächtershäuser (1988, 1992), who has argued that life began with the appearance of an autocatalytic, two-dimensional, **chemolithotrophic** (chemically powered) metabolic system based on the formation of the highly insoluble mineral pyrite (FeS_2).

Formation of pyrite by the reaction $FeS + H_2S \rightarrow FeS_2 + H_2$ is very favorable. This reaction is irreversible and highly **exergonic** (energy-liberating), producing a standard free-energy change ($\Delta G°$) of 9.23 kcal mol^{-1} that corresponds to a reduction in chemical potential ($E°$) of the reactants of 620 mV. Thus, the FeS/H_2S combination is a strong reducing agent; even under mild conditions, it should serve as an efficient source of electrons for the reduction (hydrogenation) of organic compounds. The scenario proposed by Wächtershäuser (1988, 1992) fits well with the environmental conditions found at the deep-sea hydrothermal vents, where H_2S, CO_2, and CO can be quite abundant. Interestingly, however, the FeS/H_2S system does not reduce CO_2 to amino acids, purines, or pyrimidines, although the formal chemistry of the reaction shows that the available free energy is more than adequate (Keefe et al. 1995). But pyrite formation can produce molecular hydrogen (H_2) and promote the reduction of nitrate to ammonia ($NO_3^- \rightarrow NH_3$), acetylene to ethylene ($C_2H_2 \rightarrow C_2H_4$), and thioacetic acid to acetic acid ($C_2H_4OS \rightarrow C_2H_4O_2$), as summarized by Maden (1995). Recent experiments have shown that the activation of amino acids with CO and an iron-nickel sulfide mineral ([Fe,Ni]S) can lead to formation of peptide bonds like those in proteins (Huber and Wächtershäuser 1998). (In these experiments, the required reactions take place in an aqueous environment to which the powdered sulfide mineral has been added. Thus, the reacting molecules do not form a dense monolayer bound ionically to the mineral. In contrast, the reactions discussed by Maden take place on the surfaces of pyrite grains.)

None of the FeS/H_2S system experiments shows that the postulated surface-bound metabolism gives rise to enzymes or nucleic acids. In fact, the results to date do not necessarily point to an autotrophic origin of life; rather, they are consistent with a more general, modified model of the primitive broth in which pyrite formation is an important source of electrons for the reduction of organic compounds. But possibilities remain. Under certain geological conditions (not yet described), the FeS/H_2S combination might be capable of reducing not only CO but also CO_2 (released from molten magma in deep-sea vents), thereby producing biochemically relevant monomers (Orgel 1988). Similarly, peptide polymers might be synthesized in an iron-nickel sulfide system at submarine fumaroles (Huber and Wächtershäuser 1998) if amino acids were formed earlier by Strecker-type syntheses near the ocean surface—although such a scenario would require transport of the participating amino acids from their site of formation to the deep-sea vents (Rode 1999). However, if the compounds synthesized by such processes do *not* remain bound to the pyritic surface, but drift away into the surrounding oceanic waters, they would cease to be a two-dimensional, chemically powered autotrophic organism. Instead, they would become additional components of the prebiotic soup. In sum, the experimental results achieved so far with the FeS/H$_2$S combination appear to be fully consistent with a heterotrophic origin of life.

HAVE TOO MANY COOKS SPOILED THE SOUP?

In addition to the small monomeric compounds discussed here, other kinds of organic compounds have been synthesized under primitive Earth conditions. These include dicarboxylic and tricarboxylic acids, fatty acids, and fatty alcohols. But for a variety of biochemical monomers, no syntheses under plausible prebiotic conditions have been demonstrated. Among these are arginine, lysine, histidine, thiamine, folic acid, lipoic acid, biotin, and pyridoxal. It is possible that some of these compounds were not, in fact, synthesized on the prebiotic Earth, and that their presence in living systems is a result of early intracellular metabolic evolution.

Given adequate expertise and the right experimental conditions, it is theoretically possible to synthesize almost any organic molecule. So the nonbiological synthesis of a wide array of the monomeric components of modern cells under laboratory conditions does not mean that these components were essential for the origin of life; nor does it mean that they were necessarily present in the prebiotic environment. The primordial broth may have been a chemical wonderland, but it could not have included all the compounds and molecular structures found in the simplest of present-day microbes.

The fundamental tenet informing the heterotrophic theory of life's origin is that maintenance and reproduction of the first living systems depended on prebiotically synthesized organic molecules. As summarized here, discussions about how the primitive broth was formed are both lively and numerous. Have too many cooks spoiled the soup? Certainly not. Indeed, it is very unlikely that any single mechanism can account for the wide range of organic compounds that must have been

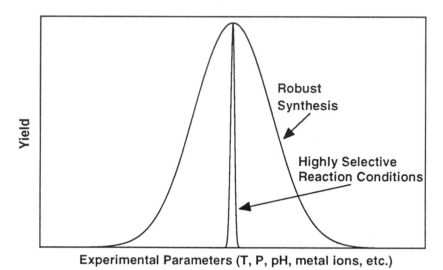

Experimental Parameters (T, P, pH, metal ions, etc.)

Figure 3.9. Unlike chemical reactions that take place under highly selective conditions, a robust organic synthesis—like that of monomers formed under possible prebiotic conditions—produces appreciable yields over a wide range of environmental settings.

present in the primitive environment. Rather, the prebiotic soup was almost certainly formed by contributions from endogenous *and* exogenous sources—from "homegrown" syntheses in a reducing atmosphere and on metal sulfides at deep-sea vents, as well as from comets, meteorites, and interplanetary dust. This eclectic view does not beg the issue of the relative significance of the different contributors of organic compounds: It simply recognizes the wide variety of potential sources of the raw materials required for the emergence of life.

The existence of diverse nonbiological mechanisms by which biochemically relevant monomers can be synthesized under plausibly prebiotic conditions is now established. Of course, not all prebiotic pathways are equally efficient, but the wide range of experimental conditions under which organic compounds can be synthesized demonstrates that prebiotic syntheses of the building blocks of life are robust. The abiotic reactions producing such compounds span a broad range of settings; they are not limited to a narrow spectrum of highly selective reaction conditions (figure 3.9). Our ideas on the prebiotic formation of organic compounds are largely based on experiments in model systems. The robustness of this approach is supported by the fact that carbonaceous meteorites contain most of the same kinds of compounds that are produced in the laboratory. Such comparisons indicate that it is certainly plausible that similar nonbiological organic syntheses took place on the primitive Earth. Yet for all the uncertainties surrounding the emergence of life, the formation of a primordial organic soup is one of the most firmly established events in Earth's history.

ACKNOWLEDGMENTS

For grant support, we thank the NASA Specialized Center of Research and Training (S. L. M.) and Proyecto UNAM-DGAPA PAPIIT-IN213598 (A. L.).

REFERENCES

Anders, E. 1989. Pre-biotic organic matter from comets and asteroids. *Nature* 342:255–57.

Baudish, O. 1913. Über nitrat- und nitritatassimilation. *Zeitschrift für Angewandte Chemie* 26:612–13.

Chyba, C. F. 1990. Impact delivery and erosion of planetary oceans in the early inner Solar System. *Nature* 343:129–33.

Chyba, C. F., and C. Sagan. 1992. Endogenous production, exogenous delivery, and impact-shock synthesis of organic compounds: An inventory for the origin of life. *Nature* 355:125–32.

Chyba, C. F., P. J. Thomas, L. Brookshaw, and C. Sagan. 1990. Cometary delivery of organic molecules to the early Earth. *Science* 249:366–73.

Clemett, S. J., C. R. Maechling, R. N. Zare, P. D. Swan, and R. M. Walker. 1993. Identification of complex aromatic molecules in individual interplanetary dust particles. *Science* 262:721–25.

Corliss, J. B., J. Dymond, L. I. Gordon, J. M. Edmond, R. P. von Herzen, R. D. Ballard, K. Green, D. Williams, A. Bainbridge, K. Crane, and T. H. van Andel. 1979. Submarine thermal springs on the Galapagos Rift. *Science* 203:1073–83.

Corliss, J. B., J. A. Baross, and S. E. Hoffman. 1981. An hypothesis concerning the relationship between submarine hot springs and the origin of life on Earth. *Oceanologica Acta* 4(Suppl.): 59–69.

Decker, P., H. Schweer, and R. Pohlmann. 1982. Identification of formose sugars, presumable prebiotic metabolites, using capillary gas chromatography/gas chromatography-mass spectrometry of *n*-butoxime trifluoroacetates on OV-225. *Journal of Chromatography* 225:281–91.

De Duve, C. 1991. *Blueprint for a cell: The nature and origin of life.* Burlington, N. C.: Neil Patterson.

Edmond, J. M., K. L. Von Damn, R. E. McDuff, and C. I. Measures. 1982. Chemistry of hot springs on the east Pacific Rise and their effluent dispersal. *Nature* 297:187–91.

Ferris, J. P., and L. E. Orgel. 1966. An unusual photochemical rearrangement in the synthesis of adenine from hydrogen cyanide. *Journal of the American Chemical Society* 88:1074.

Ferris, J. P., R. P. Sanchez, and L. E. Orgel. 1968. Studies in prebiotic synthesis. III. Synthesis of pyrimidines from cyanoacetylene and cyanate. *Journal of Molecular Biology* 33:693–704.

Ferris, J. P., O. S. Zamek, A. M. Altbuch, and H. Freiman. 1974. Chemical evolution XVIII. Synthesis of pyrimidines from guanidine and cyanoacetaldehyde. *Journal of Molecular Evolution* 3:301–9.

Ferris, J. P., P. D. Joshi, E. H. Edelson, and J. G. Lawless. 1978. HCN: A plausible source of purines, pyrimidines, and amino acids on the primitive Earth. *Journal of Molecular Evolution* 11:293–311.

Gabel, N. W., and C. Ponnamperuma. 1967. Model for the origin of monosaccharides. *Nature* 216:453–55.

Galtier, N., N. Tourasse, and M. Gouy. 1999. A nonhyperthermophilic common ancestor to extant life forms. *Science* 283:220–21.

Gibson, E. K., Jr. 1992. Volatiles in interplanetary dust particles: A review. *Journal of Geophysical Research* 97:3865–75.

Gilbert, W. 1986. The RNA world. *Nature* 319:618.

Glavin, D. P., and J. L. Bada. 1999. The sublimation and survival of amino acids and nucleobases in the Murchison meteorite during a simulated atmospheric entry heating event. *Abstracts of the 12th International Conference on the Origin of Life*, July 11–16, 1999, International Society for the Study of the Origin of Life San Diego, California, p. 108.

Glocker, G., and S. C. Lind. 1939. *The electrochemistry of gases and other dielectrics*. New York: Wiley.

Harvey, R. B. 1924. Enzymes of thermal algae. *Science* 60:481–82.

Huber, C., and G. Wächtershäuser. 1998. Peptides by activation of amino acids with CO on (Ni, Fe)S surfaces and implications for the origin of life. *Science* 281:670–72.

Kamminga, H. 1991. The origin of life on Earth: Theory, history, and method. *Uroboros* 1:95–110.

Kasting, J. F. 1993. Earth's early atmosphere. *Science* 259:920–26.

Kauffman, S. 1993. *The origins of order: Self-organization and selection in evolution*. New York: Oxford University Press.

Keefe, A. D., S. L. Miller, G. McDonald, and J. L. Bada. 1995. Investigation of the prebiotic synthesis of amino acids and RNA bases from CO_2 using FeS/H_2S as a reducing agent. *Proceedings of the National Academy of Sciences USA* 92:11904–6.

Larralde, R., M. P. Robertson, and S. L. Miller. 1995. Rates of decomposition of ribose and other sugars: Implications for chemical evolution. *Proceedings of the National Academy of Sciences USA* 92:8158–60.

Lazcano, A. 1995. A. I. Oparin: The man and his theory. In *Evolutionary biochemistry and related areas of physicochemical biology*, ed. B. F. Plogazov, B. I. Kurganov, M. S. Kritsky, and K. L. Gladilin, 49–56. Moscow: A. N. Bach Institute of Biochemistry and ANKO Press.

Lazcano, A., and S. L. Miller. 1996. The origin and early evolution of life: Prebiotic chemistry, the pre-RNA world, and time. *Cell* 85:793–98.

Leicester, H. M. 1974. *Development of biochemical concepts from ancient to modern times*. Cambridge: Harvard University Press.

Letham, D. S., and P. R. Stewart. 1977. RNA in retrospect. In *The Ribonucleic Acids*, ed. D. S. Letham and P. R. Stewart, 1–8. New York: Springer.

Levy, M., and S. L. Miller. 1999. The prebiotic synthesis of modified purines and their potential role in the RNA world. *Journal of Molecular Evolution* 48:631–37.

Levy, M., S. L. Miller, and J. Oró. 1999. Production of guanine from NH_4CN polymerizations. *Journal of Molecular Evolution* 49:165–68.

Levy, M., S. L. Miller, K. Brinton, and J. L. Bada. 2000. Prebiotic synthesis of adenine and amino acids under Europa-like conditions. *Icarus* 145:609–13.

Love, S. G., and D. E. Brownlee. 1993. A direct measurement of the terrestrial accretion rate of cosmic dust. *Science* 262:550–53.

Maden, B. E. H. 1995. No soup for starters? Autotrophy and origins of metabolism. *Trends in Biochemical Sciences* 20:337–41.

Miller, S. L. 1953. A production of amino acids under possible primitive Earth conditions. *Science* 117:528.

———. 1955. Production of some organic compounds under possible primitive Earth conditions. *Journal of the American Chemical Society* 77:2351–61.

———. 1998. The endogenous synthesis of organic compounds. In *The molecular origins of life: Assembling pieces of the puzzle*, ed. A. Brack, 59–85. Cambridge: Cambridge University Press.

Miller, S. L., and A. Lazcano. 1995. The origin of life—Did it occur at high temperatures? *Journal of Molecular Evolution* 41:689–92.

Mukhin, L. M., M. V. Gerasimov, and E. N. Safonova. 1989. Origin of precursors of organic molecules during evaporation of meteorites and mafic terrestrial rocks. *Nature* 340:46–48.

Muller, D., S. Pitsch, A. Kittaka, E. Wagner, C. E. Wintner, and A. Eschenmoser. 1990. Chemie von alpha-Aminonitrilen (135). *Helvetica Chimica Acta* 73:1410–68.

Oparin, A. I. 1924. *Proiskhozhedenie Zhizni* (Moskva: Moskovskii Rabochii). Reprinted and translated by J. D. Bernal. 1967. *The origin of life*, 199–234. London: Weidenfeld and Nicolson.

Oparin, A. I. 1938. *The origin of life*. New York: Macmillan.

Orgel, L. E. 1998. The origin of life—A review of facts and speculations. *Trends in Biochemical Science* 23:491–95.

Oró, J. 1960. Synthesis of adenine from ammonium cyanide. *Biochemical and Biophysical Research Communications* 2:407–12.

———. 1961. Comets and the formation of biochemical compounds on the primitive Earth. *Nature* 190:442–43.

Oró, J., and A. P. Kimball. 1961. Synthesis of purines under primitive Earth conditions. I. Adenine from hydrogen cyanide. *Archives of Biochemistry and Biophysics* 94:221–27.

Oró, J., and A. Lazcano. 1997. Comets and the origin and evolution of life. In *Comets and the origin and evolution of life*, ed. P. J. Thomas, C. F. Chyba, and C. P. McKay, 3–27. New York: Springer.

Pace, N. R. 1991. Origin of life—Facing up to the physical setting. *Cell* 65:531–33.

Peltzer, E. T., J. L. Bada, G. Schlesinger, and S. L. Miller. 1984. The chemical conditions on the parent body of the Murchison meteorite: Some conclusions based on amino-, hydroxy-, and dicarboxylic acids. *Advances in Space Research* 4:69–74.

Pitsch, S., R. Krishnamurthy, M. Bolli, S. Wendeborn, A. Holzner, M. Minton, C. Leseur, I. Schlönvogt, B. Jaun, and A. Eschenmoser. 1995. Pyranosyl-RNA (pRNA): Base-pairing selectivity and potential to replicate. *Helvetica Chimica Acta* 78:1621–35.

Reid, C., and L. E. Orgel. 1967. Synthesis of sugar in potentially prebiotic conditions. *Nature* 216:455.

Robertson, M. P., and S. L. Miller. 1995a. An efficient prebiotic synthesis of cytosine and uracil. *Nature* 375:772–74.

———. 1995b. Prebiotic synthesis of 5-substituted uracils: A bridge between the RNA world and the DNA-protein world. *Science* 268: 702–5.

Robertson, M. P., M. Levy, and S. L. Miller. 1996. Prebiotic synthesis of diaminopyrimidine and thiocytosine. *Journal of Molecular Evolution* 43:543–50.

Rode, B. M. 1999. Peptides and the origin of life. *Peptides* 20:773–86.

Sanchez, R. A., J. P. Ferris, and L. E. Orgel. 1966a. Conditions for purine synthesis: Did prebiotic synthesis occur at low temperatures? *Science* 153:72–73.

———. 1966b. Cyanoacetylene in prebiotic synthesis. *Science* 154:784–85.

———. 1967. Studies in prebiotic synthesis. II. Synthesis of purine precursors and amino acids from aqueous hydrogen cyanide. *Journal of Molecular Biology* 30:223–53.

———. 1968. Studies in prebiotic synthesis. IV. The conversion of 4-aminoimidazole-5-carbonitrile derivatives to purines. *Journal of Molecular Evolution* 38:121–28.

Shapiro, R. 1988. Prebiotic ribose synthesis: A critical analysis. *Origins of Life and Evolution of the Biosphere* 18:71–85.

———. 1995. The prebiotic role of adenine: A critical analysis. *Origins of Life and Evolution of the Biosphere* 25:83–98.

Stetter, K. O. 1994. The lesson of archaebacteria. In *Early life on Earth: Nobel Symposium No. 84,* ed. S. Bengtson, 114–22. New York: Columbia University Press.

Urey, H. C. 1952. On the early chemical history of the Earth and the origin of life. *Proceedings of the National Academy of Sciences USA* 38:351–63.

Vallentyne, J. R. 1963. Biogeochemistry of organic matter. II: Thermal reaction kinetics and transformation products of amino compounds. *Geochimica Cosmochimica Acta* 28:157–88.

Voet, A. B., and A. W. Schwartz. 1983. Prebiotic adenine synthesis from HCN: Evidence for a newly discovered major pathway. *Bioorganic Chemistry* 12:8–17.

Wächtershäuser, G. 1988. Before enzymes and templates: Theory of surface metabolism. *Microbiological Reviews* 52:452–84.

———. 1992. Groundworks for an evolutionary biochemistry: The iron-sulphur world. *Progress in Biophysical Molecular Biology* 58:85–201.

Zhao, M., and J. L. Bada. 1989. Extraterrestrial amino acids in Cretaceous/Tertiary boundary sediments at Stevns Klint, Denmark. *Nature* 339:463–65.

CHAPTER 4

From Building Blocks to the Polymers of Life

JAMES P. FERRIS

THE BIOPOLYMERS IN THE FIRST LIFE

Nucleic acids and **proteins** play a central role in life on Earth today. These polymeric biochemicals, composed, respectively, of **nucleotides** (figure 4.1a) and **amino acids** (figure 4.2a), provide the catalysis, the genetics, and some of the structure of all living systems. The genetic information in the nucleic acid **DNA** (deoxyribonucleic acid) is transcribed to the nucleic acid **RNA** (ribonucleic acid; figure 4.1b), and this information is then translated from RNA to protein. Most contemporary living systems need other polymers as well. Today, for example, the up-to-date organism uses polymers of sugar—**carbohydrates**—to store energy and build (plant) cell walls. It is unlikely that the first forms of life needed such polymers. But it is fairly certain that polymeric nucleotides and polymers of amino acids—or biochemicals similar to them—have been present in living systems since life's emergence.

How many different kinds of **biopolymers** were assembled in the first forms of life? It would be easier to agree on these necessary components if scientists could agree on what properties the first life forms had, or even on what "life" is (Luisi 1998). Three definitions illustrate the varying views. Version 1 is a minimalist definition, according to which life is a self-sufficient system maintained by replication and subject to change by mutation. Thus, the simplest form of life is an assemblage of nucleic acids sustained by an external source of nutrients (to provide energy), its integrity preserved by the binding of biopolymers to a mineral surface. Version 2, a more complex definition, posits a **semipermeable** barrier (a

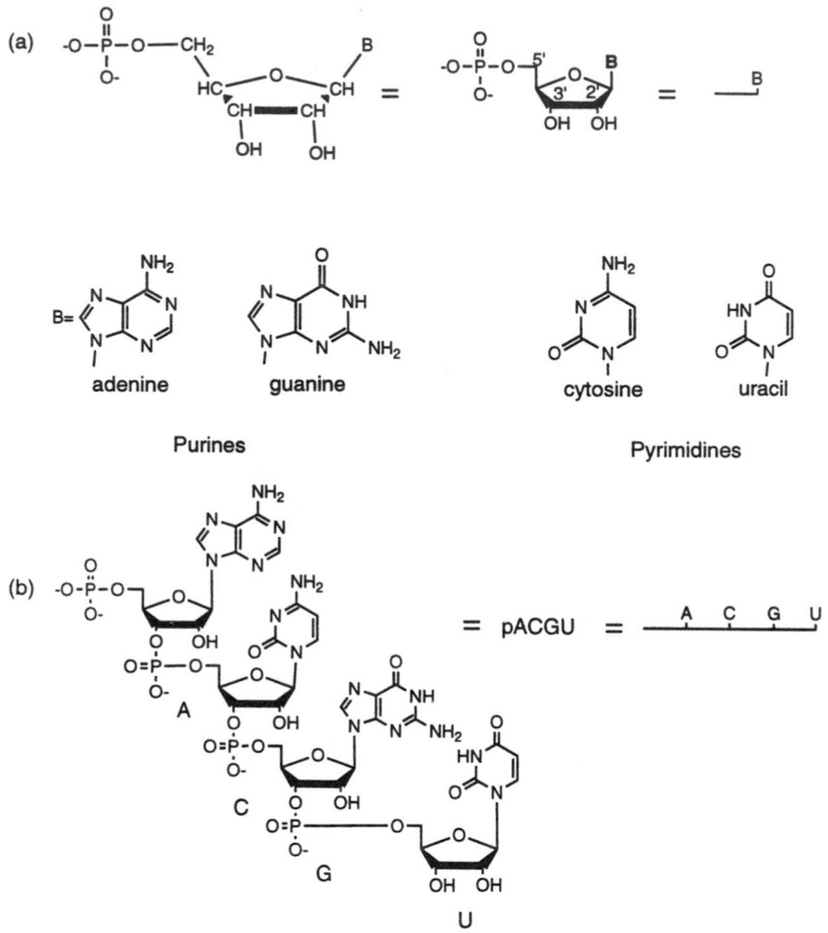

Figure 4.1. The structure of RNA. (a) The monomeric unit (a nucleotide) of RNA, showing the four nitrogenous bases (the purines, adenine and guanine; and the pyrimidines, cytosine and uracil) that can be bound into RNA nucleotides. (b) A short piece of an RNA molecule. pACGU denotes an RNA that contains the bases adenine (A), cytosine (C), guanine (G), and uracil (U).

membrane) to maintain the integrity of a primitive living system. Version 3 is more complex yet. It requires the presence of the molecular machinery (proteins) to metabolize ingested nutrients. In this case, the monomers essential to the processes of living are the products of metabolism; they are not provided by some external source.

Under these definitions, which kinds of biopolymers were required to form the first life? Version 1 requires only RNA (figure 4.1b) or its

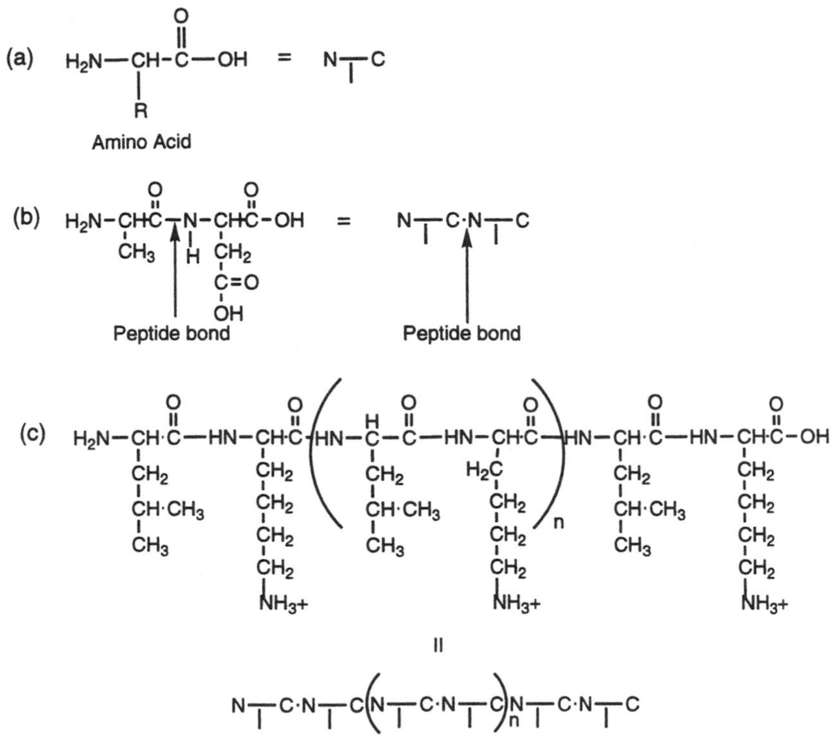

Figure 4.2. The structures of (a) an amino acid, (b) two amino acids linked by a peptide bond, and (c) a polypeptide consisting of the amino acids leucine and lysine.

equivalent, since RNA both stores genetic information and catalyzes reactions (as in **ribozymes,** a particular type of RNA). Genetic material, such as RNA or something similar, would have been essential because genetic information assures the continued existence of life and directs the formation of proteins. Version 2 requires only RNA and a semipermeable barrier (and, perhaps, **polypeptide** catalysts to manufacture the enclosing barrier). If the membrane enclosing the life forms was composed of **lipids** (as in present-day microbes), it may have required embedded proteins to control the migration of nutrients. Version 3 requires proteins to catalyze metabolism—since it is unlikely that primitive RNA or early ribozymes had the catalytic capability to carry out all of the necessary chemical transformations—as well as the complex machinery needed to make other biochemicals.

Here, we will focus on Version 2, the "middle of the road" view, offering an origin-of-life scenario that requires prebiotic synthesis of

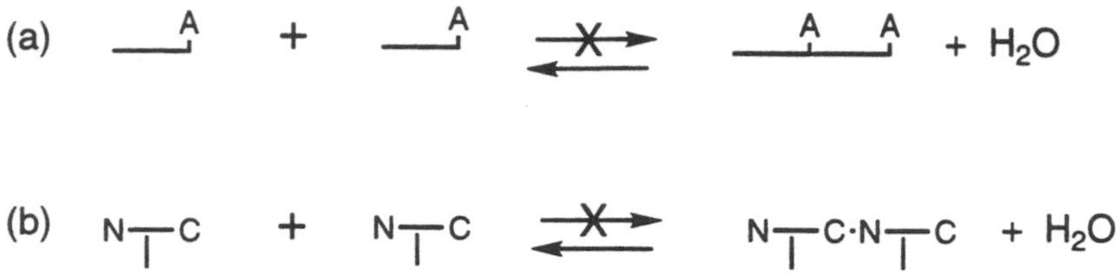

(a) ___A + ___A ⇄×→ ___A___A + H_2O
       | | ← | |

(b) N—C + N—C ⇄×→ N—C·N—C + H_2O
       | | ← | |

Figure 4.3. Formation and hydrolysis of polymer-forming chemical bonds in water. (a) Bonds linking nucleotides, as in RNA. (b) Bonds linking amino acids (peptide bonds), as in proteins. In both examples, the direct formation of bonds combining two monomers into a dimer—dinucleotides in (a), dipeptides in (b)—is not energetically favored in water; however, hydrolytic breakdown of such bonds is energetically favored.

RNA and at least some catalytic amino acid polymers. But in adopting this approach, we are forced to make a major assumption: namely, that RNA and protein (or chemicals very similar to each) were components of the first living systems. In truth, compounds quite different from RNA and proteins—for example, chemical evolutionary precursors of RNA and proteins—could have been used. But because we have few specific examples of plausible **pre-RNA** or **pre-protein** molecules, we will concentrate on nucleic acids and proteins.

RNA and proteins are polymers made from repeating monomeric subunits. In the RNA polymer (figure 4.1), the principal monomer units are derivatives of **ribose phosphate,** and each of these subunits is bound to one of four heterocyclic, nitrogenous bases: the two purines, **adenine** (A) and **guanine** (G), and the two pyrimidines, **uracil** (U) and **cytosine** (C). Proteins are polymers of amino acids (figure 4.2c) linked by peptide bonds (figure 4.2b). In contemporary life, proteins contain 20 different kinds of amino acids, each having a different side chain, or R group (shown in figure 4.2a).

General Considerations of Polymer Synthesis

Formally, the synthesis of nucleotides and amino acid polymers should be simple: Two monomers can be linked by eliminating a hydrogen atom (H) from one monomer and a hydroxyl group (–OH) from another, then releasing the resulting water molecule (HOH). But the formation of **polynucleotides** (such as RNA) or amino acid polymers (proteins) from their monomer components is not energetically favored (figure 4.3). In proteins, the polymer-forming peptide bond is thermodynamically

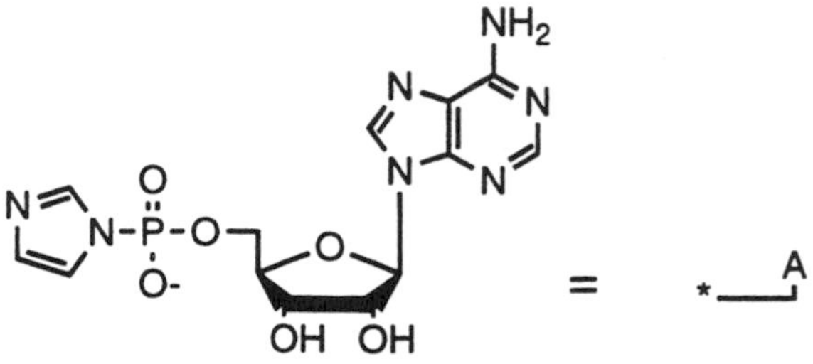

An Activated Nucleotide

ImpA

Figure 4.4. The structure of an activated nucleotide (primed for ready chemical reaction); the activating group is an imidazole group (the nitrogen-containing five-membered ring compound attached to phosphorus).

unstable in water, so in aqueous solution, energy is required to link two amino acids. Indeed, ~3.8 kcal/mol is released upon **hydrolysis** (breakdown in water) of a peptide bond, while 5.8 kcal/mol is released upon hydrolytic breakdown of a simple **amide bond** of an amino acid (Dobry and Sturtevandt 1952). And in the nucleotide 5'-AMP, hydrolysis of the **phosphate ester bond** (the kind present in all nucleic acids) releases 2.2 kcal/mol. So, on the prebiotic Earth, energy input must have been necessary to make polymers of nucleotides and amino acids.

Initial studies using heat to link monomeric units, some carried out in the absence of water, have been marginally successful. However, water was ubiquitous on the primitive Earth, so the biologically important polymers were probably formed in the presence of water. It is imperative, then, to prepare polymers in aqueous solution. One way to do that is to attach **activating groups** to the reacting monomers; such chemical groups promote linkages between monomers, priming them for reaction (figure 4.4). Another way is to use chemical **condensing agents** that provide the energy needed for bond formation.

Life on the Rocks

Even when activated monomers are used, it is difficult to form long polymers in water. Such polymers are formed in a stepwise process; and unless some additional impetus promotes polymer formation, long

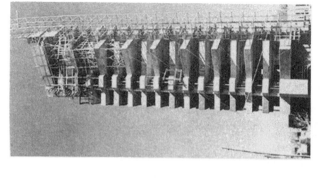

Figure 4.5. Construction of the Price Tower in Bartlesville, Oklahoma. The prebiotic formation of a polymer is analogous to constructing a tall building. Each floor must be built before the floor above it can be constructed, and many building materials are needed to complete the intermediate floors before adding the top floor. Similarly, the prebiotic formation of a long polymer requires the earlier formation of many copies of its constituent polymeric units. And such synthesis requires many building blocks (monomers) to build up the intermediate polymers before the final long polymeric chain can be made. The photos shown were prepared by Garrett Hamlin from photographs in *The Story of the Tower* by Frank Lloyd Wright, and are used with permission of Joe D. Price.

polymers can form only if an array of intermediates is formed first. For example, a 50-monomer polymer (a 50-mer) can form only if the reaction mixture contains appreciable amounts of the intermediate 2- to 49-mers (figure 4.5). To facilitate the formation of long polymers, it is desirable to limit the synthetic steps, possibly by adding activated monomers only to fairly long oligomers. One can, for example, form polymers on a mineral surface that absorbs longer oligomers from solution. But even in such systems, it remains necessary to prepare appreciable amounts of each intermediate compound.

Chemical reactions on mineral surfaces may have provided a prebiotic route to the kinds of polymers required for the first life on Earth (Bernal 1949). Polymers can bind to such substrates—longer polymers more strongly than the shorter. Leslie Orgel, who dubbed this notion **"life on the rocks,"** developed the theory for their formation (Gibbs et al. 1980; Ferris et al. 1996; Orgel 1998). Orgel noted that when a polymer binds to a mineral surface, the free energy of binding increases linearly as monomers are added, one by one, to the polymer chain. Thus, as the number of monomers increases from n to $n + 1$ units, the increase in free energy is proportional to the ratio of the distribution coefficients (K values), where

$$\Delta(\Delta F) = \frac{\left(\dfrac{\text{concentration of } n\text{-mers in solution}}{\text{concentration of } n\text{-mers adsorbed}} \right)}{\left(\dfrac{\text{concentration of } n + 1\text{-mers in solution}}{\text{concentration of } n + 1\text{-mers adsorbed}} \right)}$$

By measuring the distribution coefficients for oligoglutamic acid polymers bound to the mineral **hydroxyapatite** (figure 4.6), it is possible to determine that $\Delta(\Delta F)$, the change in free energy as each monomer unit is added, amounts to −0.85 kcal per unit. Using this value of $\Delta(\Delta F)$, it becomes possible to calculate that the distribution constant of the various oligoglutamic acid polymers on hydroxyapatite increases by a factor of 4.2 as a chain is extended by one monomer unit. A polymer made up of six glutamic acid monomers, a **hexamer** of glutamic acid (figure 4.6; $n = 4$), has a K value of ~ 1 (Hill et al. 1998). From these observations, it can be calculated that a polymer elongated to a 16-mer (figure 4.6; $n = 14$) has a distribution coefficient of about 1×10^6—a K value indicating that the polymer is permanently bound to the mineral.

The chain length achieved during the nonbiological synthesis of polymers depends on the competing rates of polymer formation and polymer

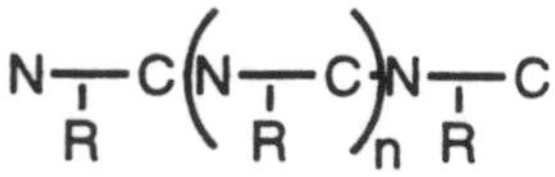

$$N\!-\!C\!\!\left(\!N\!-\!C\!\right)_{\!n}\!\!N\!-\!C$$
$$RRR$$

Figure 4.6. The structure of polyglutamic acid, a polypeptide synthesized on the surface of the mineral hydroxyapatite (R = CH_2CH_2COOH; n = 6–15).

hydrolysis. For polypeptides and nucleic acids, hydrolytic breakdown is energetically favored, but polymer formation on mineral substrates does occur. For example, the half-time for hydrolysis of the peptide bond between two glycine amino acid monomers is 500 years at 25°C (Radzicka and Wolfenden 1996); thus, a 50-monomer glycine polymer can form on a mineral substrate in 0.2 years (73 days).

From these considerations, we can draw a general conclusion: In the presence of water, only polymers capable of (1) undergoing slow hydrolysis and (2) binding to minerals can be expected to grow into long polymers under plausibly prebiotic conditions. And polynucleotides and polypeptides are two such polymers.

THE RNA WORLD: THE PREBIOTIC SYNTHESIS OF RNA

The consensus among scientists concerned with the origins of life is that RNA was the most important biopolymer in early (but not necessarily the first) life on Earth. This agreement is based on the following considerations:

1. RNA can store genetic information in its sequence of chemical bases (the pattern of distribution of its component **purines** and **pyrimidines**).

2. RNA catalyzes the transformations of RNAs and other biologically crucial compounds (for example, the synthesis of proteins, catalyzed by RNAs present in **ribosomes,** the tiny globular protein factories of cells).

3. Chemically manufactured RNAs having no catalytic activity, or only one particular catalytic function, can "evolve" under test-tube conditions into molecular structures having different catalytic functions.

Many workers studying the origin of life think that simple, precursor RNAs formed first and that modern-day RNAs evolved from these pre-RNAs. They support this idea by observing that no one has discovered the pathways to potential prebiotic sources of activated RNA monomers (for a review, see Schwartz 1998). Nevertheless, the possible steps in pre-biotic RNA synthesis are better known than are those of the proposed pre-RNAs. Of the various routes thus far explored, RNA synthesis by polymerization of activated monomers has proved the most successful.

The Montmorillonite Clay–Catalyzed Synthesis of RNA

The possible role of minerals (specifically, clay minerals) in the prebiotic reactions of organic compounds was first proposed by Bernal (1949), who focused on the selective concentration of organics by adsorption on clay surfaces, the protection of the adsorbed organics from solar radiation, and the photopolymerization of the clay-adsorbed organics.

One of the most successful applications of Bernal's proposal was the formation of RNA oligomers containing 6 to 14 monomer units—polymers formed by binding to and polymerizing on **montmorillonite** clay. In these polymers, the activating group was an **imidazole** (figure 4.4) attached to a phosphate group (Ferris and Ertem 1992; Ferris 1993). And the reaction is general: Oligomers form whether the nucleotide base involved is adenine, cytosine, guanine, or uracil (figure 4.7). Purines have also been used successfully as activating groups, but **pyrophosphates** (like the triphosphates used in living systems today) have not proved effective.

Catalysis is particularly important during such polymer formation because it has the potential to limit the number of isomers formed. A catalytic binding site can constrain the orientation of the reactants while they are undergoing chemical transitions. For example, a $3',5'$-linked phosphodiester bond should be favored in the reaction of purine nucleotides on montmorillonite. However, an alternative linkage, a $2',5'$-linked **phosphodiester bond,** is favored both in the clay-catalyzed reactions of pyrimidine nucleotides and in the absence of catalysis. (For the $2'$, $3'$, and $5'$ positions on nucleotides, and the structures of purines and pyrimidines in RNA, see figure 4.1a.)

Dimers are not formed at random. Rather, the **dimers** made in reaction mixtures containing two or more kinds of activated nucleotides show sequence selectivity (Ertem and Ferris 2000). At the end of the two-part chain produced in this first step of polymerization, formation of a $5'$-purine–pyrimidine sequence is favored over that of a $5'$-pyrimidine–purine

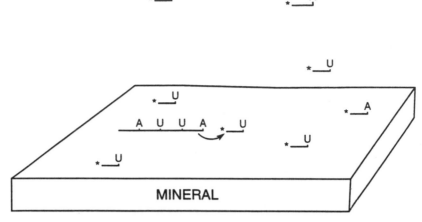

Figure 4.7. Activated monomers of the nucleotides of adenine (A) and uracil (U) reacting to form an RNA polymer on a mineral surface. In water, some activated monomers remain in solution while others become bound to the mineral surface. Elongation of pAUUA (an RNA whose nucleotides contain, sequentially, adenine, uracil, uracil, and adenine) is shown by the arrow; this indicates formation of a chemical bond from the terminal A of pAUUA to an activated monomer of U, thereby forming an RNA molecule having the composition pAUUAU.

TABLE 4.1 MAJOR DIMERIC PRODUCTS OF FOUR
ACTIVATED MONOMERS OF MONONUCLEOTIDES IN
THE PRESENCE OF MONTMORILLONITE

Dimer	Yield (%)
pApC	23
pApU	14
pGpC	14
pApA	13
pGpA	9
Total yield	73

sequence by a factor of about 20:1. We can see the selectivity imposed by montmorillonite clay if we look at the dimeric products of a reaction mixture containing the activated monomers of all four nucleotides. Although 16 dimers were possible, just five (AC, AU, GC, AA, and GA) were major reaction products, comprising 73% of the total (table 4.1). And of the 11 minor dimers (AG, CC, CU, CA, CG, UC, UU, UA, UG, GU, GG), three made up less than 2%.

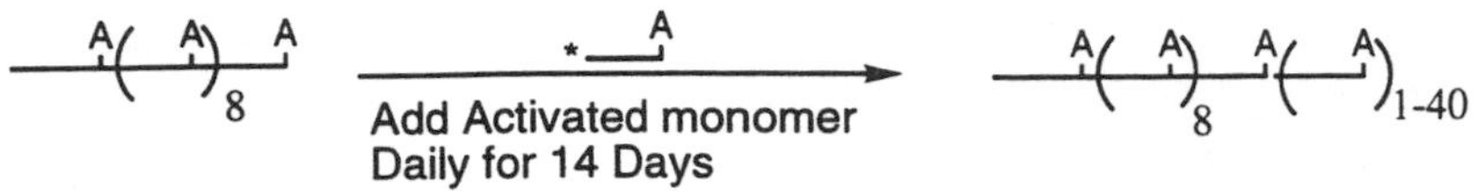

Figure 4.8. The elongation of a 10-mer made up of nucleotides of adenine bound to the clay mineral montmorillonite. The 10-mer is "fed" daily by the addition of fresh activated monomer, and the chain grows from a 10-mer to a mixture of 10-mers to 50-mers over a 14-day period.

RNA Elongation

Probably, the formation of short RNA oligomers by catalysis on montmorillonite was not sufficient to initiate the **RNA World** (discussed in chapter 5). But it is possible to use the life-on-the-rocks approach to generate fairly long RNAs by elongating short oligomers. In one set of experiments, reactive monomers were added daily to 10-mers of the nucleotide **adenosine** bound to montmorillonite, and the progress of elongation was followed for 14 days (Ferris et al. 1996). The length of the polymers increased incrementally up to 40-mers and 50-mers (figure 4.8). Comparably long oligomers of nucleotides of uracil, and of a copolymer of nucleotides of adenine and uracil, were formed in similar experiments.

On the early Earth, elongation of RNA to chains longer than 40-mers could have provided the RNAs needed to begin the RNA World. As outlined in chapter 5, it has been proposed that such RNAs would have been able to replicate by **template-directed synthesis** with sufficient fidelity to maintain in their sequence structure the overall information content needed for life (Joyce and Orgel 1999). In addition, it has been postulated that a 40-mer chain length is the minimum required for an RNA able to catalyze reactions of other RNA molecules (Joyce and Orgel 1999; Szostak and Ellington 1993). Elongation of oligomers of RNA to 40-mers or longer, now shown experimentally, may be the "missing link," bridging the gap between short polymers of RNA and a population of RNAs long enough to initiate the RNA World.

Template-Directed Synthesis

The sequence information in RNA—the pattern of distribution of its four component nucleotides—is preserved when one RNA chain serves as a template for the synthesis of its complementary chain. Then, the newly made complementary chain can be used as a template for further synthesis, resulting in formation of an RNA containing a replica of the original sequence. Nonenzymatic template-directed syntheses have used $3',5'$-linked templates, and recent research has explored the use of templates produced by catalysis on montmorillonite. In particular, Ertem and Ferris (1996) have shown that clay catalysis of activated monomers of cytosine nucleotides produces oligocytosines (oligo-Cs) having mostly (70%) $2',5'$-linkages and relatively few (20%) $3',5'$-linkages. In addition to phosphodiester bonds, the oligo-Cs thus formed contained pyrophosphate linkages, and some of the chains cyclized to make cyclic oligomers. Surprisingly, the heterogeneity of this mixture, with its diverse linkages and bond types, did not promote formation of structures that inhibit template-directed synthesis, but complementary strands of oligoguanosines (oligo-Gs) were formed when heterogeneous oligo-Cs were used as a template. Of particular interest is the observation that complementary oligo-Gs formed despite removal of the $3',5'$-linkages from the heterogeneous templates.

These studies show that the sequence information present in RNA oligomers formed by clay mineral catalysis could have been maintained by template-directed synthesis. Indeed, the diversity of the bonding present in the highly heterogeneous oligo-C mixture suggests that the sequence information present in almost any mix of RNA oligomers formed by prebiotic processes could have been maintained by such template-directed synthesis.

Metal Ion Catalysis of RNA Synthesis

Like minerals, metal ions may have played important catalytic roles in the emergence of life on Earth. Various metal ions are known to catalyze the synthesis of short oligomers of RNA from activated monomers, but the uranyl ion (UO_2^{2+}) is one of the few metal ions known to catalyze the formation of oligomers containing ten or more monomer units (Sawai et al. 1989). The phosphate-activating group used in such studies was imidazole, the same chemical group used in montmorillonite-catalyzed reactions. Oligomers of adenine, uracil, and cytosine were formed in which the maximum chain lengths obtained were 16, 12, and 10, respectively (Sawai et al. 1992). Because the uranyl ion did not catalyze the hydrolysis of the activated monomers, the extent of conversion of monomer to oligomer was very high, and most of the phosphodiester bonds in the RNAs produced were $2',5'$-linkages.

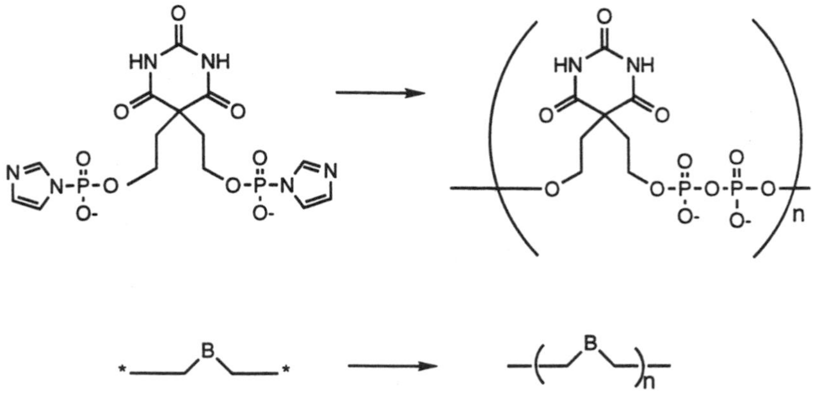

Figure 4.9. The conversion of a proposed pre-RNA monomer to a polymer whose component monomers are linked by pyrophosphate groups. (The scheme at the bottom is a simplified representation of this conversion.)

RNA Analogs

Arguing that RNA is too complicated to have formed on the primitive Earth, many scientists have begun looking for simpler structures—pre-RNAs—that could have evolved into RNA. Among these is a promising monomeric structure that may have formed oligomers containing ten or more monomer units linked by pyrophosphate groups.

Pyrophosphate groups form easily under prebiotic conditions, so their possible role in linking nucleotides has been studied in some detail. This approach was applied to a nonchiral nucleotide analog that contains a **barbituric acid** ring (figure 4.9). It is possible (but not demonstrated) that this monomer may have formed on the primitive Earth (Van Vliet et al. 1995). The barbituric acid–containing monomer is also attractive because it is not chiral; thus it is not subject to enantiomeric inhibition of polymer synthesis and does not form polymers containing both enantiomers. In the presence of a divalent manganese (Mn^{2+}) catalyst, the reaction of this monomer, both at 0°C for 14 weeks and at 37°C for 3 days, resulted in the formation of 15-mers. Other pre-RNAs also have been proposed (including some that have been used in template-directed synthesis of RNA, as discussed in chapter 5), but their plausible prebiotic syntheses have not been described.

THE ROLE OF POLYPEPTIDES IN THE ORIGIN OF LIFE

Despite many studies of the prebiotic synthesis of amino acid polymers, the relevance of such peptides to life's beginnings is not clear. How, for instance, could a complex polypeptide having a specific property (say, the ability to catalyze phosphodiester bond formation) form more than once in a purely random prebiotic polymerization process? This seems most unlikely, with a single exception—a polypeptide formed by a catalytic process in which the catalyst dictated the amino acid sequence. (In this situation, the catalyst would contain the "memory" needed to synthesize the particular sequence of amino acids in the peptide; thus, it could catalyze formation of additional copies having its particular catalytic activity.)

In the following sections, we describe the prebiotic formation of polypeptides containing more than 8 to 10 amino acids, a lower limit chosen because 8 to 10 amino acids is the optimal length of a **leucine–lysine** polypeptide chain (shown in figure 4.2c) needed to catalyze the hydrolysis of polyadenosine (poly-A).

Prebiotic Polypeptides: Structure

Several studies of well-characterized, laboratory-synthesized polypeptides offer insight into their secondary (folded) structures and catalytic properties. Today, we can tailor-make polypeptides that fold preferentially into α-helices (figure 4.10), β-sheets (figure 4.11), or random coils. And by investigating the properties of such structures, we can make informed conjectures about which types played a catalytic role in the origin of life. The β-sheet secondary structure forms spontaneously from polypeptide chains made up of amino acids with alternating hydrophobic (water-hating) and hydrophilic (water-loving) side chains, as in the leucine–lysine polymer shown in figure 4.2c (Brack and Orgel 1975). This polypeptide chain folds spontaneously into a β-sheet structure that has the hydrophilic, polar (charged) lysine–NH_3^+ side chains arrayed on one side of the sheet, and the hydrophobic (nonpolar, noncharged) leucine side chains $[-CH_2CH(CH_3)_2]$ on the other (Brack 1993, 1994).

Experimental studies support the proposal that such β-sheets played an important role in the origin of life. Polypeptide structures of this type are ten times more resistant to hydrolytic breakdown than are peptide chains made up of random coils or those having an α-helical conformation. This difference in stability suggests that polypeptides in

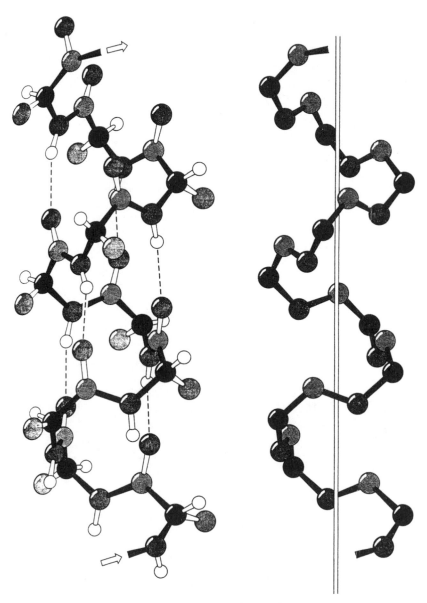

Figure 4.10. Protein-like polypeptides can fold to form α-helical structures. (Left) α-Helix with all its structural units; the dotted lines indicate intramolecular hydrogen bonds. (Right) The peptide-chain backbone of the α-helix shown, which illustrates the helical conformation of the molecule around a central axis. Reprinted from *Biochemistry* (Moran and Scrimgcour 1992), with permission from Prentice-Hall.

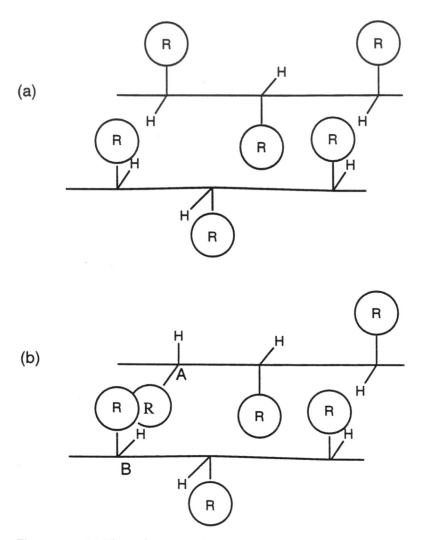

Figure 4.11. (a) The β-sheet secondary structure present in many proteins. (b) A β-sheet secondary structure in which the configuration at carbon atom A is inverted from that in normal proteins. In the latter arrangement, the R group on carbon A crowds the hydrogen attached to carbon atom B, thereby destabilizing the β-sheet.

β-sheet conformations had longer lifetimes than those in the other configurations; thus, they would have been more readily available for incorporation into the first forms of life. If so, this preference would have resulted in prebiotic chemical selection of amino acids capable of forming β-sheets. Further, such selectivity suggests that the earliest

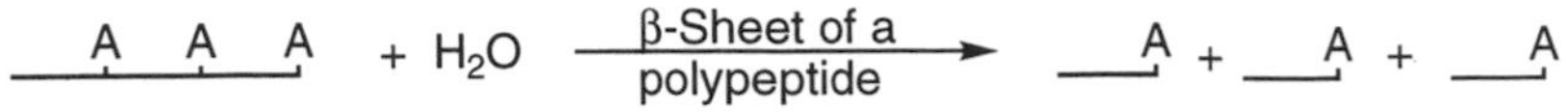

$$\underline{A \ A \ A} \ + \ H_2O \ \xrightarrow[\text{polypeptide}]{\text{β-Sheet of a}} \ \underline{A} \ + \ \underline{A} \ + \ \underline{A}$$

Figure 4.12. The hydrolytic breakdown of a polynucleotide is catalyzed by polypeptides folded into β-sheets.

biological systems contained a limited number of different amino acids (Brack 1987)—not the wide array formed in Miller-type simulation experiments (chapter 3) or brought to Earth by meteorites (chapter 2).

Prebiotic Polypeptides: Catalysis

Studies of laboratory-synthesized polypeptides also provide important clues about what type of catalytic activity the different conformations may have provided to the earliest forms of life. In particular, it has been found that multicharged (polycationic) polypeptides made up of arginine–NH_3^+ or lysine–NH_3^+ monomers and hydrophobic amino acids accelerate the hydrolytic breakdown of RNA (Barbier and Brack 1992). In these experiments, the rate of hydrolysis was greatest when the polypeptide was in the β-sheet conformation. As shown in figure 4.12, polypeptides of this type hydrolyze small RNA molecules into components that contain a repeating series of adenosines (oligo-As). The polypeptide increases the rate of hydrolysis over that observed in its absence 150-fold. The hydrolysis of RNAs having poly-U, poly-C, and poly-G compositions was also observed. And studies of the effect of polypeptide chain length on such catalytic activity reveal that more or less optimum activity is obtained with short polypeptides, those about 10 amino acids long (10-mers). Moreover, catalysis is not exclusively a feature of the β-sheet conformation, since a structurally related α-helical polypeptide [poly(leucine–lysine–lysine–leucine)] also exhibits catalytic activity, albeit at a lower (1.5- to 2.1-fold) level.

The important finding from these experiments is that short β-sheet polypeptides catalytically break down—hydrolyze—the phosphodiester bonds of RNAs. Such breakdown is just the opposite of the chemical buildup required for the origin of life (though scarcely surprising, since hydrolysis is energetically favored over phosphodiester bond formation). Yet, applied to an energetically favored reaction—such as the formation of RNAs from activated mononucleotides—this finding does suggest the possibility that a β-sheet catalyzed the prebiotic formation of RNA-type molecules.

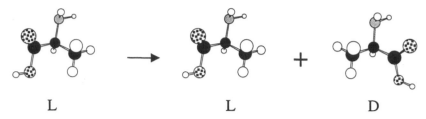

L   L   D

Figure 4.13. Amino acids in the L-configuration (such as L-alanine) can be converted into a mixture of equal amounts of L- and D-amino acids (D-alanine and L-alanine); this process is called racemization.

During laboratory syntheses of polypeptides, it is common for the component amino acids to become **racemized** (that is, amino acids of a single isomeric configuration change, so that the resulting polymer contains equal amounts of two **stereoisomers**—half in the original configuration and half in the mirror-image form). An example of the racemization of a single amino acid is shown in figure 4.13. The amino acids in living systems today are almost exclusively in the levorotatory L-**configuration** (that is, in pure solutions of such compounds, plane-polarized light is rotated to the *levo,* or leftward, direction). But laboratory experiments have shown that in partially racemized prebiotic mixtures, where some amino acids are present in both their L and D forms (pure solutions of compounds in the D-**configuration** rotate plane-polarized light in the *dextro,* or rightward, direction), the ability of a polypeptide to fold into a β-sheet structure is seriously impeded (figure 4.11a). The side chains of D-amino acids will not participate in β-sheet formation with L-amino acids because the space between the sheet-forming polypeptide chains is insufficient. (Interestingly, however, this stricture does not apply to α-helices made up of D- and L-isomers because the side chains of the amino acids are directed away from the α-helical backbone and do not impede its formation.)

Thermolysis of Amino Acids to Form Polypeptides

The initial approach to the prebiotic formation of polypeptides consisted of heating mixtures of amino acids to dryness until, with loss of a water molecule between each pair of monomers, bonds formed to generate amino acid polymers. This process did indeed produce polymers and those polymers exhibited the normal peptide bonds of proteins. However, their component amino acids were linked by other kinds of bonds as well. For example, by heating mixtures of amino acids (mixtures that typically contained a 10-fold proportion of the amino acids glutamic acid, aspartic acid, or lysine) at 180°C for 2 to 5 hours, Fox and co-workers formed globular amino acid-containing bodies dubbed proteinoids (Fox and Dose 1977). The reaction products were dark in color, suggesting that their components had at least partially decomposed. In addition, the products were linked in part by other than peptide bonds, and they were either partially or totally racemized.

From an origin-of-life point of view, the prebiotic scenario envisioned by such a process has a significant disadvantage. It requires a heating and drying of amino acids on hot volcanic rocks—a setting that may not have been widespread on the primitive Earth.

Formation of Polypeptides by Heating Amino Acid Amides

Conversion of an amino acid to its chemically activated, closely related amino acid amide (an amino acid-like compound that contains two NH_2 groups) provides sufficient energy to drive the formation of a peptide bond between the resulting amide and a nearby amino acid. Figure 4.14 shows the hydrolytic breakdown of an amino acid amide, an energetically comparable reaction that releases 5.8 kcal/mol (Dobry and Sturtevandt 1952). Thus, peptide bonds between amino acids can be formed at much lower temperatures if the starting mixture is made up of amino acid amides instead of the amino acids themselves. And, on early Earth, the amides needed to participate in such a reaction could easily have come from hydrolytic breakdown of **amino nitriles,** which are formed readily by the Strecker syntheses discussed in chapter 3.

Beginning with solutions of such amino acid amides, Ito and colleagues (1990) showed that repeated drying and heating in the presence of the clay mineral kaolin yields a substantial array of polypeptides. The presence of the clay mineral enhanced the yield of polypeptides (although the precise mechanism of this enhancement is as yet unknown). If ten hydration–dehydration cycles were used, the yield of the polymer was 10%. Heating the mixture to dryness was an essential part of the cycle; if the solutions of amino acid amides were not thus concentrated, the yields of the peptides were low. The resulting polymers were linked mainly by peptide bonds and had molecular weights between 1000 and 4000. Although some racemization of the polymer-forming amino acids occurred, the addition of divalent metal ions (Cu^{2+}, Ca^{2+}, or Zn^{2+}) to the reaction mixture suppressed it. Starting

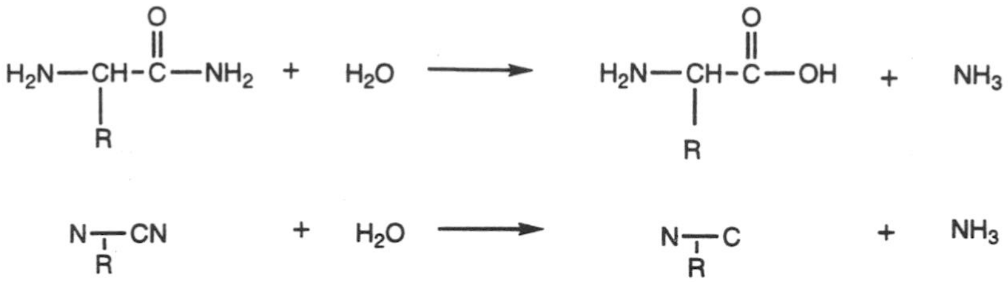

$$H_2N-CH-C(=O)-NH_2 \;+\; H_2O \;\longrightarrow\; H_2N-CH-C(=O)-OH \;+\; NH_3$$
$$\underset{R}{|} \qquad\qquad\qquad\qquad\qquad \underset{R}{|}$$

$$\underset{R}{N-CN} \;+\; H_2O \;\longrightarrow\; \underset{R}{N-C} \;+\; NH_3$$

Figure 4.14. The hydrolysis of an amino acid amide (having two NH_2 groups) to the amino acid (having one NH_2 group) and ammonia (NH_3). This reaction is energetically favored in aqueous solution, demonstrating that the amino acid amide is an activated form of the amino acid.

with mixtures of the amides of glycine, alanine, valine, and aspartic acid, Ito and colleagues used spectral analyses to show that the polymers formed contained 6% α-helices, 42% β-sheets, and 52% random conformations, and that the extent of α-helix and β-sheet formation varied with the amino acid used.

Owing to their low-temperature formation, and the resulting creation (mainly) of peptide bonds, amide-derived polypeptides have many of the properties of polypeptides found in biological systems. Importantly, amide-activated amino acids provide a route to structures whose peptide bonding is substantially or exclusively the same as that present in proteins.

Polypeptide Formation on Clay Minerals

Paecht-Horowitz and Eirich (1988) have proposed a prebiotic model based on the synthesis of proteins, as shown in figure 4.15. In this model, the aminoacylphosphate derivatives of 5'-AMP condense to polypeptides in a reaction catalyzed by montmorillonite clay. The products of this reaction are polypeptides as long as 56-mers in which the carbon-containing terminal groups are attached to the 2'- and/or the 3'-hydroxyl groups of 5'-AMP (figure 4.14).

Polymerization of N-Carboxyanhydrides

N-Carboxyanhydrides (figure 4.16) are members of a group of compounds known to chemists as Leuchs' anhydrides. They are highly effective activated amino acid derivatives known for the formation of oligopeptides in aqueous solution (Brack 1987). One possible prebiotic

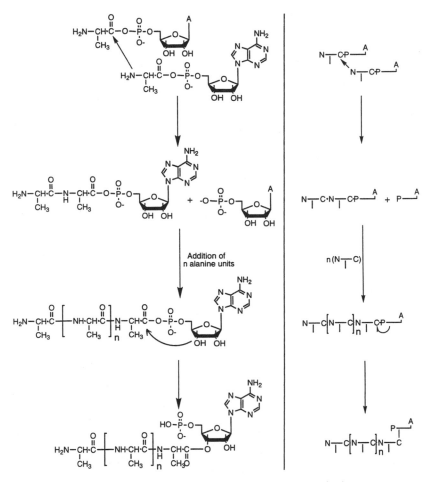

Figure 4.15. From top to bottom, a reaction sequence in which an amino acid (alanine) reacts with a nucleotide to form a polypeptide–nucleotide compound. A simplified depiction of the polymerization is shown on the right. Although reactions of this kind proceed on montmorillonite clay in the presence of water, their precise mechanisms (at the curved arrow, for example) have yet to be determined.

route to N-carboxyanhydrides is the reaction of amino acids with dicarbonyl diimidazole shown in figure 4.15a (Ehler and Orgel 1976). Peptide bond formation proceeds by the reaction of the amino group of an amino acid with the N-carboxyanhydride (figure 4.16b). N-Carboxyanhydrides of this kind are formed as reaction intermediates in aqueous solution where they rapidly polymerize to polypeptides.

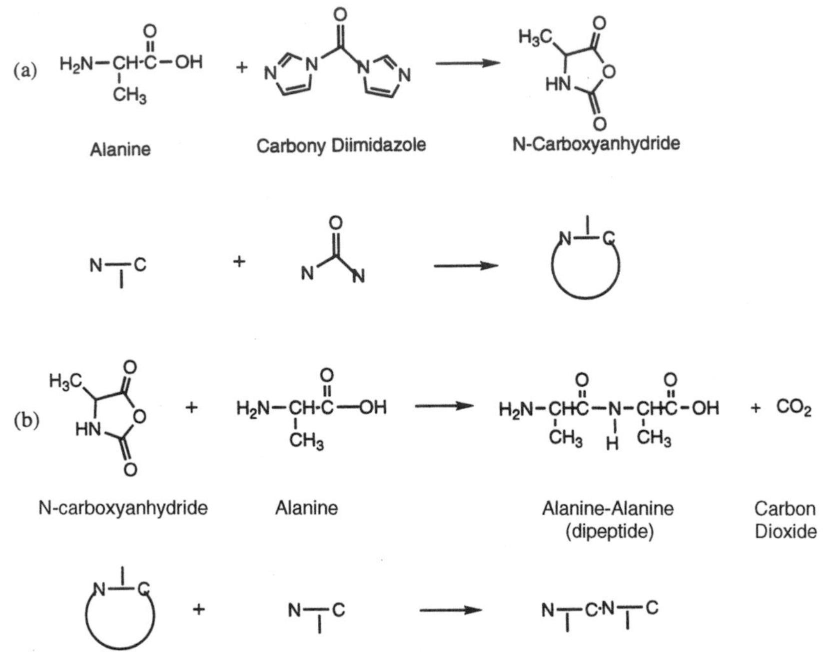

Figure 4.16. The formation of (a) an activated (*N*-carboxyanhydride) derivative of alanine by reaction with carbonyl diimidazole; (b) its subsequent reaction with alanine to form a peptide bond linking two alanine monomers into a dimer. Simplified depictions of each reaction are also shown.

The proposal that *N*-carboxyanhydrides were the precursors to polypeptides on the primitive Earth has at least five advantages:

- Under the experimental conditions (0.05–0.1 M) used to date, the rate of their polymerization in water is faster than their rate of hydrolysis.

- The amino acid components of the polymers formed are linked only by peptide bonds.

- The amino acids making up the peptides are not racemized during polymer formation.

- Oligomers containing up to 10-mers are formed in a single-step process.

- The polymerization reaction is selective for α-amino acids, the specific kind of amino acids present in the proteins of living systems.

Peptide Elongation on Minerals

The experimental procedure used to form polypeptides on mineral surfaces is generally quite similar to that described earlier for the life-on-the-rocks approach, in which polymers of RNA were successfully elongated. But if the amino acids used for the formation of polypeptides are acidic (containing two or more carboxylic acid groups and a single amino group), the process is even easier, largely because such amino acids bind tightly to a mineral substrate.

Some minerals containing divalent metal ions (such as Ca^{2+}) bind negatively charged organics to their surfaces. A good example is the mineral hydroxyapatite, $Ca_5(PO_4)_3(OH)$, in which the Ca^{2+} binds to both aspartic acid and glutamic acid (Hill et al. 1998)—the two acidic amino acids present in life today. Another mineral, the common clay mineral **illite**, also binds negatively charged amino acids (Liu and Orgel 1998). Glutamic acid oligomers longer than 45-mers were formed on hydroxyapatite and illite after 50 "feedings" of glutamic acid monomers and a chemical condensing agent (carbonyl diimidazole). Two other such experiments, using aspartic acid in place of glutamic and hydroxyapatite in place of illite, yielded polypeptides 25-mers and 13-mers in length.

CONCLUSIONS

Repeated attempts to synthesize RNA polymers by heating nonactivated monomers to dryness in the absence of catalysts have yielded only short oligomers, 10-mers or less in length. Moreover, such conditions do not mimic those prevailing on the primitive Earth. Although some high-temperature settings (for example, volcanic terrains) were certainly present, oceans and seas of liquid water dominated the planet's surface, rendering an aqueous environment a much more likely setting for prebiotic syntheses. But in an aqueous environment, RNA polymers are rapidly broken down by hydrolysis. So, assuming that the synthesis of RNA occurred at moderate temperatures in the waters of prebiotic Earth, catalyzed reactions of activated monomers must have been required. Catalysis is also required to obtain the specific (phosphodiester) bonding needed to build biologic-like RNAs and to achieve the necessary selectivity of their base sequences. Indeed, Joyce and Orgel (1999) have calculated that if two copies of all possible isomers of a 40-mer of RNA—a total of 10^{24} different isomers—were produced by a strictly random process, the mixture would have a total mass comparable to that of the entire Earth! (However, the actual outcome of such a random synthetic process using a limited number of activated monomers would be formation of oligomers much shorter than 40-mers, with very few polymers, if any, comprising as many as 50 monomer units.)

Given the difficulties encountered in experimental studies of the prebiotic formation of the activated monomers of RNA (not to mention the well documented chiral inhibition of its template-directed synthesis), many have suggested that prebiotic formation of RNAs may have been impossible on the primitive Earth. To get around this problem, numerous workers have posited the initial formation of pre-RNAs—compounds that then directed the synthesis of true RNAs. To date, however, we have seen little progress toward the synthesis of pre-RNAs. And even if such pre-RNAs were synthesized, many of the problems that confound the prebiotic synthesis of RNA itself would remain to be solved, including the problem of forming monomer subunits.

At present, the most promising approach to the prebiotic synthesis of RNA polymers involves use of the catalytic properties of mineral surfaces. In particular, it has been shown that RNA oligomers, 6-mers to 14-mers long, can be synthesized from activated monomers of adenine, cytosine, guanine, and uracil bound to the surface of montmorillonite clay. In contrast with the most successful synthesis to date of pre-RNA–type molecules (the Mn^{2+}-catalyzed formation of 15-mers of a barbituric acid derivative linked by pyrophosphate units), clay-catalyzed elongation of 10-mer primers has been shown to yield RNA-like oligomers composed of 40 or more monomer units.

Since the primitive Earth affords numerous plausible sources for amino acids, pre-amino acids have been little discussed. Attempts to synthesize polypeptides by heating mixtures of amino acids to dryness (a simulation of dying in a volcanic setting) have uniformly resulted in degradation of the amino acids and formation of polymers having bonds and other structural units that are not present in biological proteins. Currently, the most productive route to the prebiotic formation of polypeptides in aqueous solution is by means of the N-carboxyanhydride derivatives of their component amino acids.

Sequence selectivity in the prebiotic synthesis of polypeptides would be essential if these polymers were to play a useful role in the first forms of life. Before life itself evolved, no genetic system would have been available to preserve peptide sequence information (the all-important ordering of a peptide's component amino acids). So, as in the prebiotic synthesis of RNA polymers, catalysis may have been required to generate a population of peptides having the catalytic activity required for early life. But catalyzed formation of polypeptides has yet to prove fruitful. At present, therefore, the most promising solution appears to be that of generating the needed sequence selectivity of polypeptides by forming β-sheets, a conformation that undergoes hydrolytic breakdown more slowly than other conformations (although such selectivity would have been somewhat limited, because the β-sheet structure requires a polypeptide chain composed of alternating hydrophobic and hydrophilic amino acids).

ACKNOWLEDGMENTS

Support for the research described in this chapter that was carried out at Rensselaer Polytechnic Institute was provided by the National Science Foundation, the NASA Exobiology Program, and the New York Center for Studies on the Origins of Life: A NASA Specialized Center of Research and Training.

REFERENCES

Barbier, B., and A. Brack. 1992. Conformation-controlled hydrolysis of polyribonucleotides by sequential basic polypeptides. *Journal of the American Chemical Society* 114:3511–15.

Bernal, J. D. 1949. The physical basis of life. *Proceedings of the Royal Society of London* 357A:537–58.

Brack, A. 1987. Selective emergence and survival of early polypeptides in water. *Origins of Life and Evolution of the Biosphere* 17:367–79.

———. 1993. From amino acids to prebiotic active peptides: A chemical restitution. *Pure and Applied Chemistry* 65:1143–51.

———. 1994. Are peptides possible support for self-amplification of sequence information. In *Self-production of supramolecular structures. From synthetic structures to models of minimal living systems*, ed. G. R. Fleischaker, S. Colonna, and P. L. Luisi, 115–24. Dordrecht, Netherlands: Kluwer.

Brack, A., and L. E. Orgel. 1975. β-Structures of alternating polypeptides and their possible prebiotic significance. *Nature* 256:383–87.

Dobry, A., and J. M. Sturtevandt. 1952. Heat of hydrolysis of amide and peptide bonds. *Journal of Biological Chemistry* 195:141–47.

Ehler, K. W., and L. E. Orgel. 1976. N,N'-carbonyldiimidazole-induced peptide formation in aqueous solution. *Biochimica et Biophysica Acta* 434:233–43.

Ertem, G., and J. P. Ferris. 1996. Synthesis of RNA oligomers on heterogeneous templates. *Nature* 379:238–40.

Ertem, G., and J. P. Ferris. 2000. Sequence- and regio-selectivity in the montmorillonite-catalyzed synthesis of RNA. *Origins of Life and Evolution of the Biosphere* 30:411–22.

Ferris, J. P. 1993. Catalysis and prebiotic RNA synthesis. *Origins of Life and Evolution of the Biosphere* 23:307–15.

Ferris, J. P., and G. Ertem. 1992. Oligomerization reactions of ribonucleotides on montmorillonite: Reaction of the 5'-phosphorimidazolide of adenosine. *Science* 257:1387–89.

Ferris, J. P., A. R. Hill Jr., R. Liu, and L. E. Orgel. 1996. Synthesis of long prebiotic oligomers on mineral surfaces. *Nature* 381:59–61.

Fox, S. W., and K. Dose. 1977. *Molecular Evolution and the Origin of Life.* New York: Marcel Dekker.

Gibbs, D., R. Lohrmann, and L. E. Orgel. 1980. Template-directed synthesis and selective adsorption of oligoadenylates on hydroxyapatite. *Journal of Molecular Evolution* 15:347–54.

Hill, A. R. Jr., C. Böhler, and L. E. Orgel. 1998. Polymerization on the rocks: Negatively charged α-amino acids. *Origins of Life and Evolution of the Biosphere* 28:235–43.

Ito, M., N. Handa, and H. Yanagawa. 1990. Synthesis of polypeptides by microwave heating. II. Function of polypeptides synthesized during repeated hydration-dehydration cycles. *Journal of Molecular Evolution* 31:187–94.

Joyce, G. F., and L. E. Orgel. 1999. Prospects for understanding the origin of the RNA world. In *The RNA world. The nature of modern RNA suggests a prebiotic RNA,* ed. R. G. Gesteland, T. R. Cech, and J. F. Atkins. 49–77. Cold Spring Harbor, N. Y.: Cold Spring Harbor Laboratory Press.

Liu, R., and L. E. Orgel. 1998. Polymerization of β-amino acids in aqueous solution. *Origins of Life and Evolution of the Biosphere* 28:47–60.

Luisi, P. L. 1998. About various definitions of life. *Origins of Life and Evolution of the Biosphere* 28:613–22.

Orgel, L. E. 1998. Polymerization on the rocks: Theoretical introduction. *Origins of Life and Evolution of the Biosphere* 28:227–34.

Paecht-Horowitz, M., and F. R. Eirich. 1988. The polymerization of amino acid adenylates on sodium-montmorillonite with preadsorbed peptides. *Origins of Life and Evolution of the Biosphere* 18:359–87.

Radzicka, A., and R. Wolfenden. 1996. Rates of uncatalyzed peptide bond hydrolysis in neutral solution and the transition state affinities of proteases. *Journal of the American Chemical Society* 118:6105–09.

Sawai, H., K. Higa, and K. Kuroda. 1992. Synthesis of cyclic and acyclic oligocytidylates by uranyl ion catalyst in aqueous solution. *Journal of the Chemical Society, Perkin* I:505–8.

Sawai, H., K. Kuroda, and H. Hojo. 1989. Uranyl ion as a highly effective catalyst for internucleotide bond formation. *Bulletin of the Chemical Society of Japan* 62.

Schwartz, A. 1998. Origins of the RNA world. In *The molecular origins of life: Assembling pieces of the puzzle,* ed. A. Brack, 237–45. Cambridge, UK: Cambridge University Press.

Szostak, J. W., and A. D. Ellington, eds. 1993. *In Vitro selection of functional RNA sequences*. Cold Spring Harbor, NY: Cold Spring Harbor Laboratory Press.

Van Vliet, M. J., J. Visscher, and A. W. Schwartz. 1995. Hydrogen bonding in the template-directed oligomerization of a pyrimidine nucleotide analog. *Journal of Molecular Evolution* 41:257–61.

FURTHER READING

Anonymous, ed. 1987. *Evolution of catalytic function: Cold Spring Harbor Symposia on Quantitative Biology LII*. Cold Spring Harbor, N. Y.: Cold Spring Harbor Laboratory Press.

Brack, A., ed. 1998. *Molecular origins of life: Assembling pieces of the puzzle*. Cambridge: Cambridge University Press.

Fleischaker, G. R., S. Colonna, and P. L. Luisi, eds. 1994. *Self-production of supramolecular structures. From synthetic structures to models of minimal living systems*. Dordrecht, Netherlands: Kluwer.

Fry, I. 2000. *The emergence of life on Earth: A historical and scientific overview*. New Brunswick, N. J.: Rutgers University Press.

Greenberg, J. M., C. X. Mendoza-Gomez, and V. Pirronella, eds. 1993. *The chemistry of life's origin*. Dordrecht, Netherlands: Kluwer.

Schopf, J. W. 1999. *Cradle of life, The discovery of Earth's earliest fossils*. Princeton, N. J.: Princeton University Press.

Zubay, G. 2000. *Origins of life on the Earth and in the cosmos*. San Diego, Calif.: Academic.

CHAPTER 5

The Origin of Biological Information

LESLIE E. ORGEL

INTRODUCTION

Organic chemists should have invented the computer scientists' motto, "Garbage in, garbage out." Proceeding step by step, purifying the product of one step before using it in the next: this is the orthodox approach to organic synthesis. Under carefully controlled conditions, it is just possible to constrain the chemistry of a pure input compound to give a unique product. But garbage in—an impure compound or a complex mixture of compounds—does as it damn well pleases. And it almost always yields an intractable mixture of products—garbage out.

Unfortunately, all of the most impressive prebiotic syntheses produce garbage by the standards of synthetic organic chemistry. For example, Miller's classic experiment (discussed in chapter 3) produces tar along with a percent or two of a complex mixture of racemic amino acids (Miller 1953), while Oró's landmark synthesis of adenine from hydrogen cyanide (chapters 1 and 3) produces an ill-characterized brown precipitate along with, at most, a percent or two of the desired nucleoside base (Oró and Kimball 1960). How could chemistry on the primitive Earth proceed in such a messy way, producing information-rich living cells, those exquisitely designed chemical factories, from such unpromising starting materials? This is the central and as yet largely unanswered question facing investigators of the origin of life.

We know that certain polymers control the functioning of all living organisms on Earth. The **proteins,** composed of 20 standard amino acids, are the machines that catalyze almost all of the chemical reactions that go on in cells. The **nucleic acids** are largely concerned with information: **DNA** (deoxyribonucleic acid) is a blueprint that dictates which proteins are to be synthesized and when, while **RNA** (ribonucleic acid) implements DNA's instructions. And we now have overwhelming evidence (discussed in chapter 6) that all contemporary living organisms are descended from a **Last Common Ancestor** (LCA). The LCA lived on Earth at least as early as 3,500 million years ago, and the members of this plexus of early-evolved microorganisms had a biochemistry similar to that of modern bacteria. In particular, the LCA employed nucleic acids for the storage of genetic information and used proteins as catalysts. Thus, if we are to understand the origins of life, we must understand the origin of the nucleic acid–protein system.

The replication of DNA is a process brought about by protein **enzymes** called **DNA polymerases.** Provided DNA replication is sufficiently accurate, and the two DNA copies are distributed one to each daughter when a cell divides, each daughter cell starts with the same genetic information as that of the parental cell. This is the universal origin of **heredity** in biology. The DNA polymerases are proteins, the amino acid sequences of which are specified by the **base sequence** of appropriate segments of DNA. Here, we cannot go into the process of protein synthesis in detail. Suffice it to say that in the first step, the DNA sequence is **transcribed** into **messenger RNA** (mRNA) by RNA polymerases, in a process similar to that involved in DNA replication. In a second step, the RNA is **translated** into protein. The process of translation is extremely complicated, involving scores of soluble proteins and an elaborate factory known as a **ribosome.** The ribosome is made up of three RNA molecules and scores more proteins. The result of translation is the production of a protein in which the nature of each amino acid monomer in the protein sequence is determined by the nature of three **nucleotide** residues in the nucleic acid sequence. The nature of the relationship between amino acid residues and nucleotide triplets defines the **genetic code.**

This account of the genetic system, admittedly a highly oversimplified one, only emphasizes that the processes of DNA replication and protein synthesis are hopelessly intertwined in modern organisms. But even this inadequate view demonstrates a conundrum: The replication of DNA requires the pre-existence of protein enzymes, but the formation of protein enzymes requires the pre-existence of DNA. Since everyone agrees that the whole complex system could not have come into existence by chance in a single step, we have to ask the classic chicken-and-egg question: Which came first, nucleic acids or proteins?

A tentative solution emerged more than thirty years ago (Woese 1967; Crick 1968; Orgel 1968). It was suggested that a "primitive" biological system in which proteins and DNA played no part preceded the DNA–RNA–protein system. According to this hypothesis, RNA could go it alone—functioning both as a genetic material in the manner of modern DNA *and* as the basis for various types of catalysis, including the catalysis of RNA replication. These theoretical ideas received a great boost from the discovery of **ribozymes,** enzyme-like catalysts composed entirely of RNA (Kruger et al. 1982; Guerrier-Takada et al. 1983). The idea that there was once a protein-independent biological world, the so-called **RNA World,** has by now come to be widely accepted (although it remains unproven).

If we accept the RNA World hypothesis, we can restate the problem of the origin of life in a very simple two-part question: How did the RNA World come into existence, and how did this world invent the DNA–RNA–protein world? We can say very little in answer to the second part of the question, so from now on let's concentrate on the first part, the origin and development of the hypothesized RNA World.

THE RNA WORLD

It would be a mistake to think of the RNA World as fixed and unchanging. Presumably, it started simple, but it finished complicated enough to invent protein synthesis. Very different opinions about the complexity of the RNA World have been expressed, probably because some workers focus on its primitive beginnings while others concentrate on its mature state, just prior to the invention of ribosomal protein synthesis. In the following, we first consider a very optimistic scenario that explains how a simple RNA World could get started under very favorable conditions—conditions admittedly different from those prevailing on the primitive Earth. This is "The Molecular Biologist's Dream" (Joyce and Orgel 1999), which posits a primitive pool loaded with chemically activated nucleotides waiting to be polymerized into RNA. Later, we will take a more realistic view of the problem.

The Molecular Biologist's Dream

On the way to an RNA World from a pool of nucleotides pre-activated for polymerization—say as D-ribonucleoside-5'-triphosphates (figure 5.1a)—the first step must have been the formation of **polynucleotides**

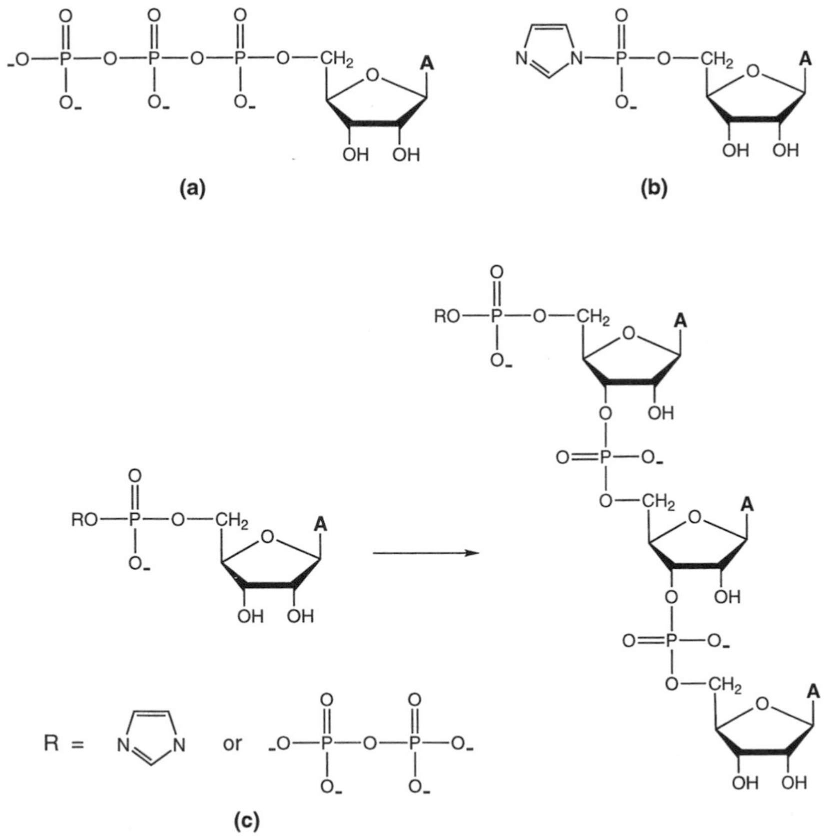

Figure 5.1. The synthesis of an oligonucleotide from an activated mononucleotide. (a) ATP, the substrate of enzymatic nucleic acid synthesis. (b) An imidazolide of a nucleotide of the kind used in many nonenzymatic template-directed reactions. (c) The synthetic reaction leading to the formation of a trinucleotide.

(figure 5.1c). Consequently, a good deal of experimental work on the nonenzymatic polymerization of activated mononucleotides has been reported (Orgel 1998). Unfortunately, nucleoside triphosphates react very slowly at moderate temperatures; this is an obstacle to laboratory investigation, although it may have been unimportant on the primitive Earth. Most recent experiments have been carried out using nucleoside-5'-phosphorimidazolides (figure 5.1b), compounds that can be considered triphosphate analogs. At first sight, the polymerization of nucleoside-5'-phosphorimidazolides doesn't seem very promising, since it yields a complex mixture of very short **oligonucleotides.** The naturally occurring oligonucleotides are hardly formed at all. As discussed in chapter 4, James Ferris and his co-workers offered a partial solution to this problem (Ferris 1998). They have shown that a common clay mineral, **montmorillonite,** is an excellent catalyst for the oligomerization of adenosine-5'-phosphorimidazolide and the corresponding derivatives of the other nucleoside bases. Oligoadenylic acids made up of as many as 40 to 50 subunits (40- to 50-mers) have been obtained in this way (Ferris et al. 1996), and nothing seems to impede the generation of **heterooligomers** containing all four bases. Furthermore, in some (but not all) cases, the products formed when montmorillonite is used as a catalyst are predominantly in the same (3'–5') linkage as in biological nucleic acids. Of course, the results obtained by Ferris and colleagues do not come close to providing the mixture of oligonucleotides needed to facilitate the next phase of the molecular biologist's dream. But they do seem to justify the hope that with some activated derivative and some catalytic mineral it might be possible to generate just such a pool of oligonucleotides. Let's assume that this is the case, and that a random mixture of oligonucleotides became available on the primitive Earth.

The development of a self-sustaining RNA World from a pool of random polynucleotides could have occurred only by Darwinian selection at the molecular level. But no matter how remarkable its properties, any single RNA molecule would have broken down in a fairly short time. The only way a molecule could have influenced the long-term history of its environment is through repeated replication. Provided new copies of a successful sequence were formed before the ancestral copy was hydrolyzed, a single ancestral sequence could have sired an ever-increasing family of descendents. And once a number of ancestral sequences with different replication rates began to compete with each other, only the most efficient replicators would have survived.

These considerations lead us to the most important recent advance in our understanding of the origin of life—the realization that the existence of polymeric molecules capable of replicating and competing is both necessary and sufficient to account for the emergence of life. Next we must ask another two-part question: Could RNA replicate without the help of protein enzymes, and do RNA molecules have attributes that permit **Darwinian evolution** via natural selection (competition)? The jury is still out, but a good deal of encouraging experimental evidence is already available.

Although the replication of nucleic acids in living organisms depends on protein enzymes, the nucleotide bases are exquisitely preadapted to

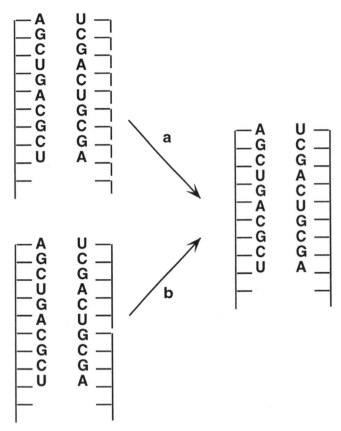

Figure 5.2. Template-directed oligonucleotide synthesis. A complete copy of the template can be obtained by (a) polymerizing monomers or (b) ligating oligomers.

serve as the components of a replicating system. For example, both the A–T (**adenine–thymine**) and G–C (**guanine–cytosine**) base pairs form spontaneously when simple N-methyl derivatives of the bases are dissolved in organic solvents (Katz et al. 1965). More importantly, monomeric derivatives of the purine bases line up spontaneously on complementary polymers of pyrimidine nucleotides at fairly low temperatures (Howard et al. 1966; Huang and Ts'o 1966); this is the prime property that makes nonenzymatic copying of nucleic acids relatively straightforward. For example, if an activated derivative containing monomers of **guanylic acid** (G) is lined up on a template made of polycytidylate (poly-C), the hydroxyl groups of each monomer are brought close to the activated phosphate groups of its neighbor, thus facilitating a rapid and efficient reaction (figure 5.2a).

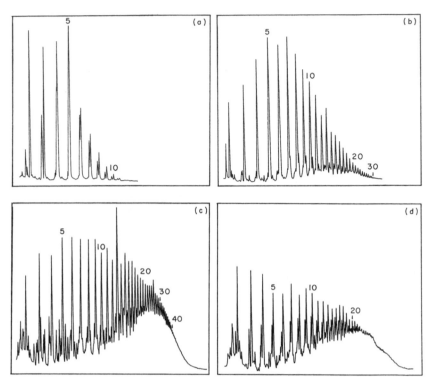

Figure 5.3. Template-directed synthesis on polycytidylate (poly-C). High-performance liquid chromatograph (HPLC) elution profiles of self-condensation of 2-MeImpG on poly-C at various pH values: (a) pH 7.0; (b) pH 7.6; (c) pH 8.2; (d) pH 9.0. The numbers above selected peaks are the lengths of the corresponding oligomers. Reaction conditions: 0°C for 14 days; 0.4 M 2,6-lutidine·HCl buffer; 0.1 M 2-MeImpG; 0.1 M poly-C; 1.2 M NaCl; 0.2 M MgCl$_2$. Conditions for HPLC: elution with a linear NaClO$_4$ gradient (pH 12, 0–0.06 M, 90 min); ultraviolet absorption monitored at 262 nm (Inoue and Orgel 1982).

Many reactions of this type have been described. Here, we concentrate on reactions carried out in our laboratory at the Salk Institute, using the 2-methyl imidazole derivatives of the nucleotides as substrates. The products formed when guanosine-5′-phosphoro-2-methylimidazolide (2-MeImpG) oligomerizes by itself in aqueous solution are relatively short and uninteresting. But when a poly-C template is present (Inoue and Orgel 1982), a remarkable reaction occurs and long oligomers, up to at least the 40-mers, can be detected (figure 5.3). This is the prototype of the template-directed reactions that we now discuss in a little more detail.

Our work on template-directed prebiotic synthesis of nucleic acids could be considered completely successful only if all possible sequences could be copied. As it turns out, this is far from the case. However, through a series of detailed experimental studies, we have shown that a great many sequences can be used as templates. The findings can be partially summarized as follows:

· Most templates containing more than 60% of C (cytosine) can be copied efficiently, but templates containing less than 60% of C cannot be copied.

· Templates made up largely of C and G (guanine) cannot be copied efficiently even if they contain an excess of C.

· Sequences of two or more A (adenine) residues in the template block elongation of the product oligomers.

Despite these drawbacks, quite long oligomers can be copied successfully. An example of this kind of reaction (Acevedo and Orgel 1987) is shown in figure 5.4.

Work in our laboratory has concentrated on the reactions of **mononucleotides,** but ligation reactions between oligonucleotides are also possible (figure 5.2b). Reactions of this kind are easier to study because they can be carried out at higher temperatures (Rohatgi et al. 1996). It is particularly promising that substrates very similar to those used in enzymatic replication **ligate** very specifically on appropriate RNA templates. It is less easy to decide whether monomers or short oligomers are the preferred substrates for simulations of prebiotic reactions. Oligomers would probably work better, but it is much harder to understand how they could have formed spontaneously on the primitive Earth (Ferris and co-workers notwithstanding).

Although much work remains to be done, the results obtained so far suggest that the formation and copying of long, single-stranded, RNA oligomers could have occurred spontaneously, given a pool of activated monomers. In many ways, the most exciting advance in prebiotic chemistry in the last decade is the demonstration that a pool of random RNA oligomers of a size that can easily be handled in the laboratory always contains substantial numbers of molecules that are enzyme-like catalysts, and that relatively simple selection procedures can be used to isolate very active catalysts from random mixtures. RNA molecules that catalyze specific chemical reactions are called ribozymes.

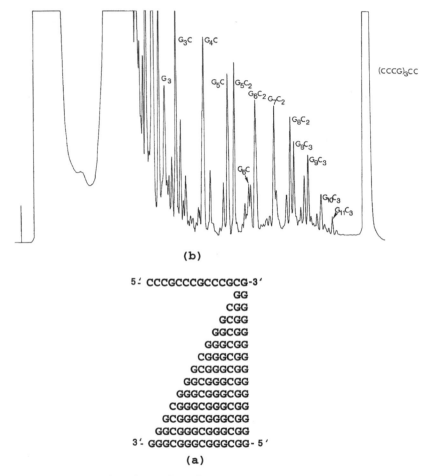

(b)

5′ CCCGCCCGCCCGCG-3′
GG
CGG
GCGG
GGCGG
GGGCGG
CGGGCGG
GCGGGCGG
GGCGGGCGG
GGGCGGGCGG
CGGGCGGGCGG
GCGGGCGGGCGG
GGCGGGCGGGCGG
3′ GGGCGGGCGGGCGG- 5′

(a)

Figure 5.4. Template-directed synthesis on a 14-mer template. (a) The expected products; (b) the identified products.

The isolation of new ribozymes is a very active field of research, but it is already clear that ribozymes can catalyze many of the steps that would be needed to create a self-sufficient RNA organism (table 5.1; Bartel and Unrau 1999). For example, a single ribozyme can catalyze both the synthesis of a nucleotide from a base and the synthesis of an activated **ribose phosphate,** while another can act as a polymerase capable of efficiently copying an RNA template up to four or five bases long. It is scarcely excessive optimism to believe that we will soon be able to isolate ribozymes capable of carrying out the same reactions that RNA and DNA polymerases perform.

TABLE 5.1 EXAMPLES OF NEW RIBOZYMES FROM
RANDOM-SEQUENCE RNA SELECTIONS

Bond Formed	Leaving Group	Activity of Ribozyme
$-O-PO_3-$	5'-RNA	Phosphodiester cleavage
$HO-PO_3-$		Cyclic phosphate hydrolysis
$-O-PO_3-$	PP_i	RNA ligation
$-O-PO_3-$	PP_i	Limited polymerization
$-O-PO_3-$	AMP	RNA ligation
$-O-PO_3-$	ADP	RNA phosphorylation
$-O-PO_3-$	Imidizole	Tetraphosphate cap formation
$-O-PO_3-$	Rpp	Phosphate anhydride transfer/hydrolysis
$-O-PO_3-$	PP_i	RNA branch formation
$-O-CO-$	AMP	RNA aminoacylation
$-O-CO-$	3'-RNA	Acyl transfer
$-O-CO-$	AMP	Acyl transfer
$-HN-CO-$	3'-RNA	Amide bond formation
$-HN-CO-$	AMP	Peptide bond formation
$-N-CH_2-$	I	RNA alkylation
$-S-CH_2-$	Br	Thioalkylation
$-HC-CH$		Diels–Alder addition (anthracene–maleimide)
$-N-CH$	PP_i	Glycosidic bond formation

The three research programs just described form the framework for achieving the molecular biologist's dream (figure 5.5). A reaction of the type studied by Ferris and co-workers could have led to the formation of a pool of random-sequence oligonucleotides, including, perhaps, an RNA polymerase. Nonenzymatic copying could have converted single-stranded RNA to double-helical RNA. Finally, the functional strand of the double helix could have copied the complementary nonfunctional strand to form a second functional strand. Then, repeated copying of functional and nonfunctional strands could have given rise to an exponentially growing population of RNA copies capable of undergoing selection, just as in the laboratory experiments on the selection of functional RNA sequences. For this scenario to work, we have to assume that the double-helical RNAs were isolated from each other to prevent a successful polymerase from copying useless or competing RNA sequences. At this point, enclosure in a membrane or some other means of isolation would have become necessary. Attachment to a

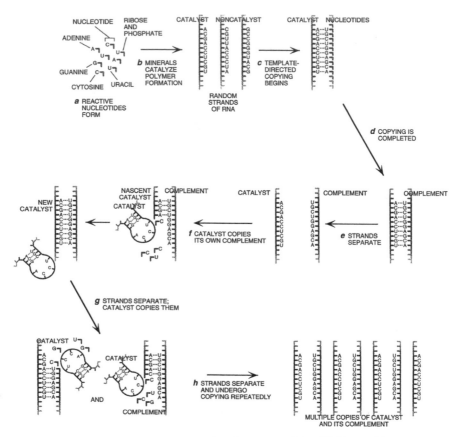

Figure 5.5. An optimistic scenario for the origin of the RNA World. (a) Pre-biotic synthesis of activated nucleotides. (b) Synthesis of random oligomers on a suitable mineral. (c, d) Nonenzymatic, template-directed synthesis of the complements of the random oligomers. (e) Chain separation, the critical step, in which one of the separated strands folds to form an RNA polymerase–like ribozyme. (f–h) Repeated rounds of ribozyme-catalyzed copying.

mineral surface or some kind of a colloidal aggregate could well have preceded enclosure in a well-defined micelle.

WHAT'S WRONG WITH
THE MOLECULAR BIOLOGIST'S DREAM?

The trouble with this dream is that nucleotides are complicated molecules. Many of the steps needed to form these nucleotides do not proceed efficiently under prebiotic conditions. For example, the prebiotic synthesis of sugars from **formaldehyde** gives a complex mixture of products, in which ribose is just a minor component. Moreover, it is difficult to form nucleosides from a base and a sugar, and the synthesis of pyrimidine nucleosides has not been achieved under plausibly prebiotic conditions. Similarly, the **phosphorylation** of nucleosides tends to give a complex mixture of products (Ferris 1987). The inhibition of the template-directed reactions on D-templates by L-substrates is a difficulty of a different, and more fundamental, kind. No one has found a way to form D-**nucleotides** without forming an almost equal amount of L-**nucleotides** (Joyce et al. 1984). It is nearly inconceivable that nucleic acid formation and replication could have begun without a much simpler mechanism for the prebiotic synthesis of nucleotides. Eschenmoser and his colleagues (Müller et al. 1990) have had considerable success in generating ribose 2,4-diphosphate in a potentially prebiotic reaction from very simple starting materials—glycoaldehyde monophosphate and formaldehyde. This suggests that the direct prebiotic synthesis of nucleotides by novel chemistry is not hopeless. Nonetheless, it seems more likely that some simpler organized form of chemistry preceded the RNA World. This leads us to a discussion of **genetic takeover.**

Genetic Takeover and Pre-RNA Worlds

Cairns-Smith has emphasized how improbable it is that a molecule as complex as RNA could have appeared de novo on the primitive Earth. To get around this problem, he proposed that the first form of life was a self-replicating clay (Cairns-Smith 1982). In this **Clay World,** the synthesis of organic molecules became part of a competitive strategy in which the inorganic clay **genome** was taken over by one of its organic creations. Cairns-Smith's postulate of a primordial inorganic life form has gathered no experimental support. However, he also contemplated the possibility that RNA was preceded by one or more simpler linear organic genomes (Cairns-Smith and Davies 1977). This idea is now very popular, but its full implications have not always been appreciated.

If RNA was not the first genetic material, the biochemistry of living organisms may not provide any clues to the earliest chemistry involved in the origin of life. The biological world that immediately preceded the RNA world must already have had the capacity to synthesize nucleotides. This should help us guess something about its chemical characteristics. However, if there were two or more "worlds" before the RNA World, the original chemistry might have left no trace in contemporary biochemistry.

In that case, the chemistry of the origins of life is unlikely to be discovered without investigating all the chemistry that might have occurred on the primitive Earth, whether or not that chemistry has any relation to biochemistry. At present we can only speculate.

The only informational systems, other than nucleic acids, that have been discovered are fairly closely related to nucleic acids. Eschenmoser and his colleagues (Eschenmoser 1997) have undertaken a systematic study of the properties of nucleic acid analogs in which either ribose is replaced by another sugar or the normal **furanose** form of ribose (a five-membered ring) is replaced by the **pyranose** form (a six-membered ring). Polynucleotides based on the pyranosyl isomer of ribose (**p-RNA**) are able to pair tightly, thus forming Watson–Crick double helices (figure 5.6), which are more stable than RNA; moreover, p-RNAs are less likely than the corresponding natural RNAs to form multiple-strand competing structures (Eschenmoser 1997). Pyranosyl RNA seems to be an excellent candidate for a genetic system; it might even be an improvement on the standard nucleic acids in some ways. More recently, Eschenmoser has reported that certain oligomers containing **tetrose** (a four-carbon sugar) rather than **pentose** (a five-carbon sugar) also form base-paired double helices (Eschenmoser 1999). However, prebiotic synthesis of either pyranosyl nucleotides or tetrose-based nucleotides does not seem much easier than synthesis of the standard isomers.

Peptide nucleic acids (PNAs), also nucleic acid analogs, have been studied extensively (figure 5.6c). PNAs were synthesized by Nielsen and colleagues (Egholm et al. 1992) in an attempt to develop inhibitors of protein synthesis (antisense RNAs). A PNA molecule is an uncharged, **achiral** analog of a standard nucleic acid in which the sugar–phosphate backbone of RNA or DNA is replaced by a backbone held together by amide bonds. PNA forms very stable double helices with complementary molecules of RNA or DNA (Egholm et al. 1992, 1993). We have shown that information can be transferred from PNA to RNA, and vice versa, in template-directed reactions (Schmidt et al. 1997a,b), and that PNA–DNA chimeras form readily on either DNA or PNA templates (Koppitz, Nielsen, and Orgel 1998). Thus, a transition from a PNA World to an RNA World is possible. Nevertheless, it is unlikely that PNA was important on the early Earth, because PNA monomers cyclize when they are activated; this would make oligomer formation very difficult under prebiotic conditions.

The studies just described suggest that there are many ways of linking nucleotide bases into chains that can form Watson–Crick double

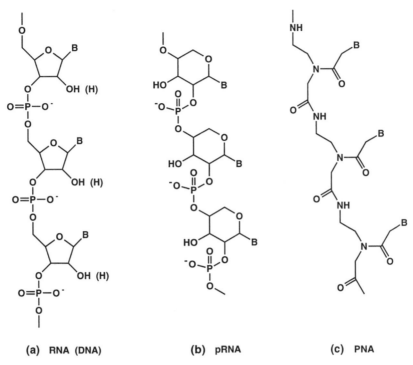

(a) RNA (DNA)   **(b) pRNA**   **(c) PNA**

Figure 5.6. Informational polymers (a) RNA (ribonucleic acid) and, denoted by parentheses, DNA (deoxyribonucleic acid). (b) p-RNA (pyranosyl RNA, a ribonucleic acid in which the five-membered furanose form of ribose is replaced by its six-membered isomer, the pyranose form). (c) PNA (peptide nucleic acid, an analog of RNA or DNA having a backbone held together by amide bonds rather than the standard sugar–phosphate linkage).

helices. If we could discover a structure of this kind that could be synthesized easily under prebiotic conditions, it would be a strong candidate for the very first genetic material. However, another possibility remains to be explored: namely, that the first genetic material might have involved no nucleoside bases at all. Two or more very simple molecules could have the pairing properties needed to form a genetic polymer—a positively charged and a negatively charged amino acid, for example. However, it is not clear that stable structures of this kind exist. Because the RNA backbone permits simultaneous base pairing *and* base stacking, the RNA molecule clearly is adapted to double-helix formation. It is unclear that molecules much simpler than a nucleotide could substitute for them to produce a stacked structure. Perhaps some other interaction between the chains can stabilize a double helix in the absence of base-stacking. Binding to a mineral surface might supply the necessary constraints, but this remains to be demonstrated. In the absence of experimental evidence, we can say little that is useful. Clearly, this is a key problem for students of the origins of life, and it should be tackled vigorously by experimentalists as soon as possible.

THE METABOLIST ARGUMENT

A very large gap separates the limited complexity of molecules that are readily synthesized in simulations of the chemistry of the early Earth from the complicated molecules known to form potentially replicating informational structures. Several workers have therefore proposed that metabolism may have come before genetics (Kauffman 1986; Wächtershäuser 1988; De Duve 1991). Their idea is that substantial organization of reaction sequences can occur in the absence of a genetic polymer and, hence, that the first genetic polymer probably appeared in an already specialized chemical or biochemical environment. Because it is hard to envisage a chemical cycle that produces β-D-nucleotides, this theory would fit best if a simpler genetic system preceded RNA.

There is no agreement concerning the extent to which metabolism could develop independently of a genetic material. As far as I am aware, no experimental evidence supports the notion that long sequences of reactions can organize spontaneously. The problem of achieving sufficient specificity, whether in aqueous solution or on the surface of a mineral, is so intractable that there is little chance of closing a cycle of reactions as complex as, say, the reverse citric acid cycle. The metabolist argument is in desperate need of experimental support.

CONCLUSION AND OUTLOOK

During the past few decades, we have learned a great deal about the origin of life. The realization that an RNA World preceded the emergence of our DNA-RNA-protein world is perhaps the most important new insight. However, a very large gap separates the prebiotic chemistry of small molecules from the chemistry of the RNA World. Presumably, the gap will begin to close as chemists and molecular biologists pursue their collaborative research programs.

ACKNOWLEDGMENTS

This work was supported by NASA (grant number NAG5-4118) and NASA NSCORT/EXOBIOLOGY (grant number NAG5-4546). I thank Aubrey R. Hill, Jr. for technical assistance and Bernice Walker for manuscript preparation.

REFERENCES

Acevedo, O. L., and L. E. Orgel. 1987. Non-enzymatic transcription of an oligodeoxynucleotide 14 residues long. *Journal of Molecular Biology* 197:187–93.

Bartel, D. P., and P. J. Unrau. 1999. Constructing an RNA world. *Trends in Biochemical Sciences* 24:M9–M12.

Cairns-Smith, A. G. 1982. *Genetic takeover and the mineral origins of life.* Cambridge: Cambridge University Press.

Cairns-Smith, A. G., and C. J. Davies. 1977. The design of novel replicating polymers. In *Encyclopaedia of Ignorance,* ed. R. Duncan and M. Weston-Smith, 391–403. New York: Pergamon.

Crick, F. H. C. 1968. The origin of the genetic code. *Journal of Molecular Biology* 38:367–79.

De Duve, C. 1991. *Blueprint for a cell: The nature and origin of life.* Burlington, N. C.: Neil Patterson.

Egholm, M., O. Buchardt, P. E. Nielsen, and R. H. Berg. 1992. Peptide nucleic acids (PNA). Oligonucleotide analogues with an achiral peptide backbone. *Journal of the American Chemical Society* 114:1895–97.

Egholm, M., O. Buchardt, L. Christensen, C. Behrens, S. M. Freier, D. A. Driver, R. H. Berg, S. K. Kim, B. Norden, and P. E. Nielsen. 1993. PNA hybridizes to complementary oligonucleotides obeying the Watson-Crick hydrogen-bonding rules. *Nature* 365:566–68.

Eschenmoser, A. 1997. Towards a chemical etiology of nucleic acid structure. *Origins of Life and Evolution of the Biosphere* 27:535–53.

———. 1999. Chemical etiology of nucleic acid structure. *Science* 284:2118–24.

Ferris, J. P. 1987. Prebiotic synthesis: Problems and challenges. In *Evolution of catalytic function, Cold Spring Harbor symposia on quantitative biology LII,* 65–83. Cold Spring Harbor, N. Y.: Cold Spring Harbor Laboratory Press.

———. 1998. Catalyzed RNA synthesis for the RNA world. In *Molecular origins of life: Assembling pieces of the puzzle,* ed. A. Brack, 255–68. Cambridge: Cambridge University Press.

Ferris, J. P., A. R. Hill, Jr., R. Liu, and L. E. Orgel. 1996. Synthesis of long prebiotic oligomers on mineral surfaces. *Nature* 381:59–61.

Guerrier-Takada, C., K. Gardiner, T. Marsh, N. Pace, and S. Altman. 1983. The RNA moiety of ribonuclease P is the catalytic subunit of the enzyme. *Cell* 35:849–57.

Howard, F. B., J. Frazier, M. F. Singer, and H. Miles. 1966. Helix formation between polyribonucleotides and purines, purine nucleosides and nucleotides. II. *Journal of Molecular Biology* 16:415–39.

Huang, W. M., and P. O. P. Ts'o. 1966. Physicochemical basis of the recognition process in nucleic acid interactions. I. Interactions of polyuridylic acid and nucleosides. *Journal of Molecular Biology* 16:523–43.

Inoue, T., and L. E. Orgel. 1982. Oligomerization of (guanosine 5'-phosphor)-2-methylimidazolide on poly(C). *Journal of Molecular Biology* 162:201–17. In *The RNA World. The nature of modern RNA suggests a prebiotic RNA,* ed. R. G. Gesteland, T. R. Cech, and J. F. Atkins, 49–77. Cold Spring Harbor, N. Y.: Cold Spring Harbor Laboratory Press.

Joyce, G. F., and L. E. Orgel. 1999. Prospects for understanding the origin of the RNA world. In *The RNA World. The nature of modern RNA suggests a prebiotic RNA,* ed. R. G. Gesteland, T. R. Cech, and J. F. Atkins, 49–77. Cold Spring Harbor, N. Y.: Cold Spring Harbor Laboratory Press.

Joyce, G. F., G. M. Visser, C. A. A. van Boeckel, J. H. van Boom, L. E. Orgel, and J. van Westrenen. 1984. Chiral selection in poly(C)-directed synthesis of oligo(G). *Nature* 310:602–4.

Katz, L., K. Tomitu, and A. Rich. 1965. The molecular structure of the crystalline complex ethyladenine: Methyl-bromouracil. *Journal of Molecular Biology* 13:340–50.

Kauffman, S. A. 1986. Autocatalytic sets of proteins. *Journal of Theoretical Biology* 119:1–24.

Koppitz, M., P. E. Nielsen, and L. E. Orgel. 1998. Formation of oligonucleotide-PNA-chimeras by template-directed ligation. *Journal of the American Chemical Society* 120:4563–69.

Kruger, K., P. J. Grabowski, A. J. Zaug, J. Sands, D. E. Gottschling, and T. H. R. Cech. 1982. Self-splicing RNA: Autoexcision and autocyclization of the ribosomal RNA intervening sequence of *Tetrahymena*. *Cell* 31:147–57.

Miller, S. L. 1953. A production of amino acids and possible primate earth conditions. *Science* 117:528–29.

Müller, D., S. Pitsch, A. Kittaka, E. Wagner, C. E. Wintner, and A. Eschenmoser. 1990. Chemie von a-Aminonitrilen. Aldomerisierung von Glycolaldehyd-phosphat zu racemischen Hexose-2,4,6-triphosphaten und (in Gegenwart von Formaldehyd) racemischen Pentose-2,4-diphosphaten: *rac*-Allose-2,4,6-triphosphat und *rac*-Ribose-2,4-diphosphat sind die Reaktionshauptprodukte. *Helvetica Chimica Acta* 73:1410–68.

Orgel, L. E. 1968. Evolution of the genetic apparatus. *Journal of Molecular Biology* 38:381–93.

———. 1998. The origin of life—a review of facts and speculations. *Trends in Biochemical Sciences* 23:491–95.

Oró, J., and A. P. Kimball. 1960. Synthesis of adenine from ammonium cyanide. *Biochimica et Biophysica Research Communications* 2:407–12.

Rohatgi, R., D. P. Bartel, and J. W. Szostak. 1996. Kinetic and mechanistic analysis of nonenzymatic, template-directed oligoribonucleotide ligation. *Journal of the American Chemical Society* 118:3332–39.

Schmidt, J. G., L. Christensen, P. E. Nielsen, and L. E. Orgel. 1997. Information transfer from DNA to peptide nucleic acids by template-directed syntheses. *Nucleic Acids Research* 25:4792–96.

Schmidt, J. G., P. E. Nielsen, and L. E. Orgel. 1997. Information transfer from peptide nucleic acids to RNA by template-directed syntheses. *Nucleic Acids Research* 25:4797–4802.

Wächtershäuser, G. 1988. Before enzymes and templates: Theory of surface metabolism. *Microbiological Reviews* 52:452–84.

Woese, C. R. 1967. *The Genetic Code*. New York: Harper.

CHAPTER 6

When Did Life Begin?

J. WILLIAM SCHOPF

INTRODUCTION

We have a fairly clear picture of *how* life began. Sketched in broad strokes, a six-part scenario is plausible: (1) the genesis in distant stars of the chemical elements crucial to life; (2) the formation of the Solar System and accretion of planet Earth; (3) the nonbiologic buildup in Earth's oceans of small, simple, organic **monomers**; (4) the linkage of these monomers into larger, more complicated, **polymeric** organics; (5) the rise of information-containing polymers; (6) the aggregation and assembly of living cells.

But if the broad strokes are clear, the details are not. We cannot, for example, estimate the likelihood of these steps. Was the origin of life rapid and "easy"—an all but inevitable byproduct of planetary forma-tion? Or, was it slow and "difficult"—a tortuous process that hinged on a highly improbable series of vanishingly unlikely chance events? It's entertaining to ponder such questions; but, lacking detailed knowledge, we cannot expect firm answers, even for a planet such as our own. And whether life's origin was likely or not, solving that puzzle would tell us nothing about what the first life forms were or when they actually arose.

Even if we cannot reason from the plausible *how* of life's beginnings to the likely *what* and *when*, we have other arrows in our quiver. Three lines of evidence can help us answer these key questions. First, we can construct a "family tree" that shows the genealogical relations among all organisms living today. Using that tree to determine which biologic traits date from near life's beginnings, we can gain insight into what the earliest life forms may have been like. Second, we can use the ancient fossil record of minute cellular organisms and their chemical signatures left in rocks. Thus, we can trace life's roots back though time to show what kinds of organisms were present early in Earth history and set a firm date on how long life has existed. Third, we can read the history of our planet's formation from the scarred and cratered surface of its Moon, thus shedding light on when Earth first became a habitable abode. Taken together, the evidence drawn from such methods reveals that life evolved very rapidly and remarkably early.

EVIDENCE FROM LIVING ORGANISMS

The Basics of Life

All living systems are made up of a small number of rather simple chemical ingredients. The most abundant and essential of these is common water (H_2O), the medium that makes up the bulk of every living system—95% of a lettuce leaf, 75% of a bacterial cell. This watery milieu harks back to life's beginnings. Whether life began in oceans, lakes, or shallow lagoons, it unquestionably got started in aqueous environs, a heritage carried over to the watery sap, the **cytosol**, of living cells where the chemistry of life takes place. And water is not the only ingredient universal to life. More than 99.9% of the substance of each living system consists of but four chemical elements—carbon, hydrogen, oxygen, and nitrogen (CHON)—elements that are among the most abundant in the cosmos. Because atoms of CHON are able to combine with one another to form small, sturdy molecules such as **methane** (CH_4), **carbon dioxide** (CO_2), and **ammonia** (NH_3), or to pair with atoms of their own kind as in gaseous **oxygen** (O_2), **hydrogen** (H_2), and **nitrogen** (N_2)—and because all of the compounds thus formed dissolve readily in H_2O (another linked pair of the prime four)—the chemistry of all of life, **biochemistry,** is largely the chemistry of CHON.

At an organismal level, life displays remarkable variety—bacteria, whales, palm trees, ants—a host of forms so different one from another that we can barely imagine commonalities among them. But at a chemical level, all organisms are nearly identical. The atoms of life are always present in the same suite of small building-block molecules or **monomers**—just three dozen or so of the same **amino acids, sugars, purines, pyrimidines,** and the like. And in all life, these monomers are linked to form the same few kinds of large **polymers, the proteins, carbohydrates,** and **nucleic acids** so important to the workings of life. The sameness carries over to **metabolism** as well, since energy to power life is produced in just a few relatively simple and closely related ways. All this sameness derives from a single evolutionary tree.

The shared basics of life demonstrate that *all* living systems—all organisms over all of time—trace their roots to the same parent cell line. The "remarkable variety" we see in the living world reflects evolutionary modifications of but a single biologic blueprint.

The Universal Tree of Life

A powerful new technique makes it possible to construct a genealogical family tree that reveals not only the relatedness of seemingly disparate organisms (bacteria, whales, palm trees, ants) but also their relatedness to the original parent cell line. This technique tests a particular biochemical present in the organisms sampled, but cannot be used on fossils (the key molecule is rapidly destroyed, even under mild conditions of geologic preservation). But because the biochemical tested is present in every living organism, the technique yields a **Universal Tree of Life.** This branching treelike diagram shows the evolutionary relations among *all* organisms living today. This new approach is possible for the following reasons.

1. Every organism manufactures proteins (polymers made up of amino acids, such as the **enzymes** that speed metabolism), which are always made in cells on the surfaces of tiny, ball-shaped protein factories called **ribosomes.**

2. The ribosomal protein factories are mostly made of various kinds of ribonucleic acid (**RNA**), long stringlike polymers having a sugar–phosphate backbone and a riblike array of four nitrogen-containing monomers: two **purines** (**adenine** and **guanine**) and two **pyrimidines** (cytosine and uracil).

3. **Mutations** (changes in the chemistry of an organism's chromosomal genes) can alter the makeup and shift the order of the purines and pyrimidines in ribosomal ribonucleic acid (**rRNA**).

4. Members of biologic groups closely related by evolution have similar genes; hence the rRNAs of closely related organisms are similar, while the rRNAs of more distantly related organisms are dissimilar.

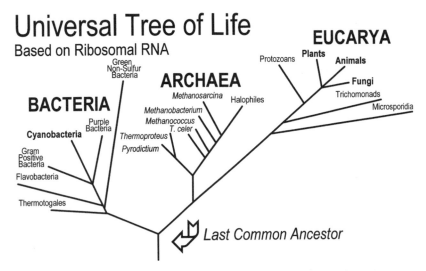

Figure 6.1. The Universal Tree of Life based on 16S ribosomal RNA.

Comparison of the patterns of distribution of purines and pyrimidines in the rRNAs in living members of the major groups of life gives a clear index of how closely or distantly related the organisms in them are. The resulting Universal Tree (shown in figure 6.1, based on a kind of rRNA known as 16S), constructed so the length of each branch corresponds to the number of changes in the pattern that happened since the branch spouted from its nearest neighbor, shows four main facts.

1.   Although plants and animals are the most familiar forms of life, they comprise just two of more than 20 major evolutionary branches.

2.   The tree is overwhelmingly composed of microscopic organisms: Of the many branches, only three (plants, animals, and fungi) contain forms large enough to see with the naked eye, and each of these contains microscopic forms as well.

3.   Every present-day organism belongs to one of three super-kingdom-like clans known as **domains** (figure 6.2): *Bacteria*— a domain comprising non-nucleated small-celled life forms, including all the common bacteria and the cyanobacteria (biochemically advanced microorganisms that give off oxygen as a byproduct of photosynthesis); *Archaea*—a domain comprising non-nucleated microbes, the archaeans, including many kinds of **extremophiles** (microorganisms able to thrive

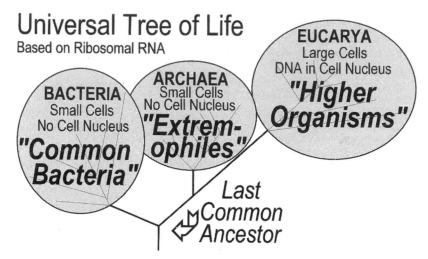

Figure 6.2. The three domains of life shown by the Universal Tree.

in exceedingly acidic high-temperature settings) and **methanogens** (microbes that give off methane gas as a product of metabolism); and *Eucarya*—the domain comprising eukaryotes (such as plants and animals), higher organisms having relatively large cells in which chromosomes are packaged in a saclike nucleus.

4. Eukaryotes are more closely related to archaeans than to bacteria, and the root of the Universal Tree—a lineage of life referred to as the Last Common Ancestor—lies between the Bacteria and the Archaea.

The branching pattern of the Universal Tree agrees fairly well with the known fossil record—Bacteria and Archaea early, Eucarya much later. But rRNA trees cannot show precisely when the various branches sprouted. Such trees are based on organisms living today; if they clocked evolution accurately, all the branches would be exactly the same length, their branchtips corresponding to the present day. But as shown in figure 6.1, this is not so. The branches are decidedly of differing lengths—some long, others short—because the different branches of life evolved at different rates. The relatively long-branched groups (for example, green non-sulfur bacteria and flavobacteria) evolved more rapidly, whereas lineages having shorter branches (several in the Archaea and eukaryotic plants, animals, and fungi) evolved more slowly.

The Last Common Ancestor

Biologic evolution is among the most conservative and economical processes imaginable. Instead of inventing new kinds of molecules, complicated biochemical systems, or novel tissues and organs, it builds bit by bit on what already exists. Moreover, because bits can be added only to what already exists, and because biochemical systems comprise many intricately interlinked pieces, any particular full-blown system can arise only once. Why the process operates this way is easy to understand. Cells are like sophisticated timepieces whose working parts depend on one another in complicated ways. Big changes are likely to destroy the delicate balance. But small changes can be tolerated—generally, the more minor, the better. Over enormous spans of time, such small changes add up; but were it not for the vastness of geologic time, the effects would be all but imperceptible.

Because of the way evolution works, the genealogical relations shown by the Universal Tree can be used to explore the nature of life's **Last Common Ancestor** (LCA)—the plexus of primitive early-evolved microbes that existed before the rise of superkingdom-like domains—even though all members of this ancient rootstock are long extinct. Since any complete biochemical system is far too elaborate to have evolved more than once in the history of life, it is safe to assume that microbes of the primal LCA cell line had the same traits that characterize all its present-day descendants. In other words, all the key components shared by every member of the three domains must have been present in LCA microbes: **genes** (DNA), proteins (including enzymes), and protein factories (ribosomes and RNA), together with the ways to make energy (metabolic pathways), store energy (**ATP, adenosine triphosphate**), repair mutations and deformed enzymes (repair mechanisms and chaperonin proteins), and divide and reproduce (by special **genome**-replicating enzymes). The basic workings of life—from genetics to the manufacture of proteins, from metabolism to the way cells divide—were in place before the ancient rootstock gave rise to present-day domains.

A second way to explore the nature of the LCA lineage is to consider traits that may be special to particularly primitive microbes. One such trait stands out (figure 6.3). The deepest branches of the Universal Tree are populated by **thermophiles**, heat-loving microbes able to thrive at high temperatures, many above the boiling point of water (possible only because the microbes live at deep-sea volcanic vents where the pressure of the overlying ocean keeps water from turning to steam).

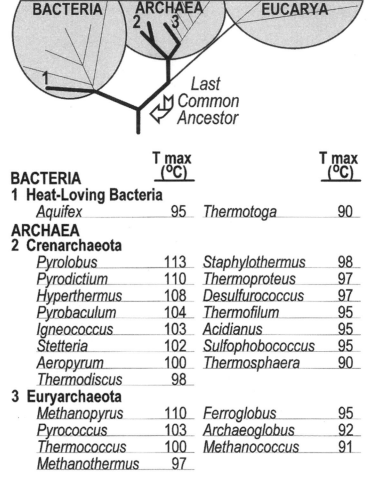

BACTERIA	T max (°C)		T max (°C)
1 Heat-Loving Bacteria			
Aquifex	95	Thermotoga	90
ARCHAEA			
2 Crenarchaeota			
Pyrolobus	113	Staphylothermus	98
Pyrodictium	110	Thermoproteus	97
Hyperthermus	108	Desulfurococcus	97
Pyrobaculum	104	Thermofilum	95
Igneococcus	103	Acidianus	95
Stetteria	102	Sulfophobococcus	95
Aeropyrum	100	Thermosphaera	90
Thermodiscus	98		
3 Euryarchaeota			
Methanopyrus	110	Ferroglobus	95
Pyrococcus	103	Archaeoglobus	92
Thermococcus	100	Methanococcus	91
Methanothermus	97		

Figure 6.3. Heat-loving microbes, both bacteria (1) and archaeans (2 and 3), populate the deepest branches of the Universal Tree.

The one-to-one match between exceptional tolerance to high temperature and extraordinary primitiveness, both in the Bacteria and the Archaea, suggests that the LCA rootstock was composed, entirely or in part, of heat-loving microbes. But it does not follow, as some workers have argued, that life itself originated at high temperatures near deep-sea vents. Even if the ancient LCA lineage was made up exclusively of heat-loving microbes, these thermophiles must have been biochemically complex, thoroughly adaptable, and highly evolved. In many respects, then, such microorganisms were practically "modern" and

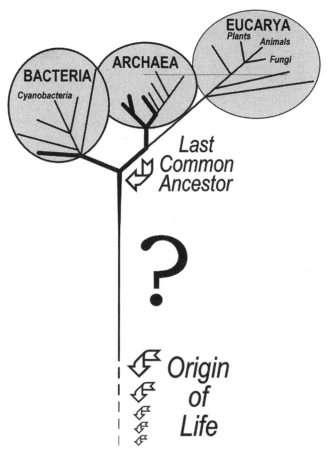

Figure 6.4. The Last Common Ancestor, the rootstock of the Universal Tree that probably included heat-loving microbes, was only distantly related to the earliest forms of life.

hence too distant from life's beginnings to reveal how living systems got started (figure 6.4).

Strengths and Weaknesses of the Universal Tree

The rRNA Universal Tree offers a new way to trace the course of biologic evolution. Truly universal, it applies to every kind of present-day organism over the entire globe. It demonstrates how simple the living world actually is, comprising three large superkingdom-like groupings instead of a multitude of fundamentally different forms. It shows the unity of life, all organisms belonging to a single evolutionary tree, all descending from the same primal rootstock, all basically alike. And it reveals the key underpinnings of evolution itself, the marked conservatism and economy of process.

But the Universal Tree is not a panacea. Sampling only present-day biologic groups, it neglects the vast majority (more than 99%) of species that have existed over life's long history. It gives a clear view of the overall branching pattern, but it gives no date for when the tree took root or when its many branches sprouted. And because it traces life's roots only to the Last Common Ancestor, when life was already highly evolved, the Universal Tree sheds no light on how life began.

EVIDENCE FROM THE ROCK RECORD

To ferret out the *what* and *when* of life's beginnings, we must turn to the geologic record, a source of *direct* evidence far closer to the events in question than the living organism–based Universal Tree.

All of geologic time, the total history of the Earth, is divided into two great eons. The older of these, the **Precambrian Eon,** extends from the planet's formation, about 4,500 million years (**Ma**) ago, to the appearance of fossils of hard-shelled animals (trilobites and mollusks) about 550 Ma ago. The younger and shorter eon, the **Phanerozoic,** spans the most recent 550 Ma and the familiar evolutionary progression: from seaweeds to marsh plants and land plants (with seeds and flowers); from animals without backbones to fish and land-dwelling vertebrates, then to birds and mammals. When we think of ancient life, we usually think of lobster-like trilobites, dinosaurs, or saber-toothed cats, the large fossils of the Phanerozoic. But the rocks of this younger eon actually record only a brief late chapter—the most recent one-eighth—of a very much longer evolutionary story. This story begins early in Precambrian time and involves microbes rather than plants and animals, that is, bacteria and archaeans rather than eukaryotes. Obviously, our search for the oldest records of life must focus on rocks of the Precambrian and, of these, on rocks dating from more than 3,000 million years ago.

From its beginnings in the early 1800s, scientific study of the Phanerozoic history of life has concentrated on fossils—their body forms, life histories, habitats, evolutionary relations, and spread in time and space. It was only in the 1960s, after more than a century of unrewarded search, that a series of breakthrough discoveries extended such studies into the Precambrian. While this long-delayed progress met with resounding success—and a stunning sevenfold increase in the doc-umented antiquity of life—investigation of the minute life forms of the Precambrian raises new challenges, demanding an approach apprecia-bly broader than that typical of traditional fossil-hunting.

Almost all Phanerozoic fossils belong to groups living today, so we rarely ask questions about their basic living processes, their **physiology** and metabolism. Even extinct forms are immune; no one wonders whether the earliest small land plants were capable of **photosynthesis** or whether dinosaurs were oxygen-breathing. Yet in the Precambrian, such questions are pivotal. Much of early evolution hinged on the develop-ment of new physiological capabilities—the origin of photosynthesis, for example, or of oxygen-breathing **aerobic respiration.** But for Precambrian microbes, the existence of such capabilities is often difficult to show by classical fossil-centered techniques, largely because microorganisms of the same size and shape can differ markedly in physiology and metabolism.

The emergent discipline of **paleobiology** has begun to answer the fresh questions raised by Precambrian life. This broad new science, which seeks evidence of the living processes of ancient life, goes beyond the fossil-focused approach. One of the key techniques now used routinely in paleobiologic studies of the Precambrian is isotopic geochemistry, a potent tool used to track the geologic history of photosynthesis.

Isotopic Evidence of Photosynthesis

Isotopes are forms of an element that differ slightly in atomic mass but have almost the same chemical behavior. Some isotopes decay radio-actively; others are immutable. Carbon has isotopes of both kinds. Carbon-14 (^{14}C) decays so rapidly that after 60,000 years or so, it be-comes undetectable (limiting the use of carbon-14 dating to prehistoric remains younger than this age). In addition, carbon has two immutable isotopes: the common ^{12}C and the slightly heavier ^{13}C. Preserved in sed-imentary rocks, these can be traced back through geologic time to pro-vide evidence of ancient photosynthetic life.

Photosynthesis is the life-sustaining biologic process characteristic of land plants, seaweeds, and some microbes (cyanobacteria and various kinds of photosynthetic bacteria). In this process the energy of sunlight is used to link molecules of carbon dioxide (CO_2) to atoms of hydro-gen to form **glucose** sugar ($C_6H_{12}O_6$)—a critical biochemical because its breakdown provides C, H, and O for use in building other bio-chemicals while fueling further photosynthetic growth. The isotopic

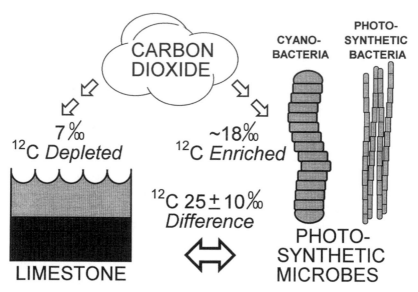

Figure 6.5. Microbial photosynthesis partitions carbon isotopes between limestone and photosynthetic bacteria and cyanobacteria.

makeup of the glucose formed, and hence of the entire organism, is controlled by the mix of isotopes in the gases and simple nutrients taken in from the environment. For the stable isotopes of carbon, this is a mixture of two kinds of carbon dioxide: $^{12}CO_2$ and $^{13}CO_2$.

In the first step of photosynthesis, a special enzyme captures CO_2 and brings it together with hydrogen to produce a small CHO-containing compound, which is converted to glucose by further steps. But $^{12}CO_2$ and $^{13}CO_2$ have different weights so they are ensnared by the enzyme with different ease. Molecules of the two types of carbon dioxide bounce about in the cell sap like tiny ping-pong balls—the lighter $^{12}CO_2$ rapidly, the heavier $^{13}CO_2$ more slowly. The lighter molecules hit the enzyme more frequently, so they are caught and bound into the initial product more often, creating a disparity that extends to glucose, the source of carbon for all other organic compounds of the organism. As a result of this enzyme-guided process, the mix of atoms in the organic matter of photosynthesizers contains more ^{12}C, and correspondingly less ^{13}C, than the mixture in the atmospheric CO_2.

Carbon dioxide is also captured in minerals. When CO_2 dissolves in seawater, it is converted into **bicarbonate** (HCO_3^-). Bicarbonate then combines with dissolved calcium (Ca^{2+}) to make mineral grains of **calcium carbonate** ($CaCO_3$), which rain down onto the ocean floor to

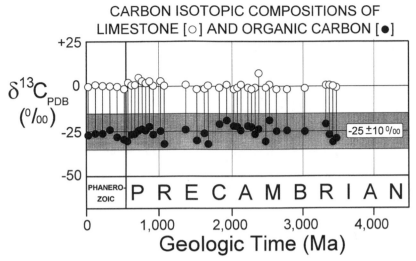

Figure 6.6. The carbon isotopic signature of photosynthesis extends to 3,500 Ma ago.

form **limestone.** This process separates the stable isotopes of carbon, but the effect is the opposite of that produced by photosynthesis—the isotopic mix in limestone contains less ^{12}C, and more ^{13}C, than the mixture in the CO_2 of the atmosphere.

Compared with atmospheric CO_2, photosynthetic microbes (like all photosynthesizers) are enriched in ^{12}C, whereas limestones are depleted (figure 6.5). The differences are small but distinct, a few percent or less, measured by **mass spectrometry** in parts per thousand (permil, ‰). For purposes of comparison, measurements in laboratories worldwide are calibrated against a standard specimen (the PDB limestone) whose assigned value is 0.0‰. Differences from this standard are expressed as $\delta^{13}C_{PDB}$ values (in which δ denotes difference). By this yardstick, samples containing relatively more of the heavier ^{13}C have positive values and those having more ^{12}C are negative.

The distribution of carbon isotopes leaves a telltale imprint in the rock record that can be used to track photosynthesis through geologic time. Compared to the CO_2 carbon source, photosynthetic microbes are enriched in the lighter isotope, generally by about 18‰, and limestones are depleted by 7‰ (figure 6.5). The two diverge by about 25‰, a clear difference that can be measured easily in geologic samples. In fact, as shown in figure 6.6, this signature of photosynthesis can be traced far into the geologic past. From the present back to

~3,500 Ma ago, we see a steady clustering of $\delta^{13}C_{PDB}$ values, with fossil organic matter hovering near $-25‰$ and ancient limestones near 0‰. And evidence suggests that this record may extend even further. For example, crystalline **graphite** from 3,800-Ma–old rocks of southwestern Greenland (sediments of the Isua Supracrustal Group, the oldest strata known in the geologic record) has an isotopic composition that hints at the presence of photosynthesis.

From the evidence of carbon isotopes, we can see that the biochemically complicated process of photosynthesis had already evolved as early as 3,500 Ma ago. Moreover, the story read from carbon isotopes is backed by fossils—minute, cellularly preserved members of the bacterial domain that similarly date from about 3,500 Ma ago.

Fossil Evidence of Ancient Microbes

As a result of ordinary geologic processes, all rock units have a limited lifetime. Typically, they are deposited, then uplifted in massive mountains, where they are eroded by wind and rain. Weathered away bit by bit, their mineral grains are carried by streams and rivers until they are finally laid down as part of some newly forming rock unit. Because of this geologic recycling, the hunt for records of ancient life, in the form of telltale isotopic signatures or in cellularly preserved fossils, becomes more and more chancy as the search moves into ever-older terrains. Rocks surviving to the present become increasingly rare as the rock record gradually peters out (figure 6.7); moreover, the few truly ancient units that do survive are often so severely pressure-cooked that evidence of life is obliterated or blurred beyond recognition. The severely heated rocks of the Isua Sequence of Greenland are a case in point. Although it is certainly imaginable that life existed 3,800 Ma ago when these rocks were laid down, nothing remains today except chemical hints—and *only* hints—of its possible presence.

In addition to the rocks of southwestern Greenland, we know of only two exceptionally ancient terrains, both made up of strata 3,000 to 3,500 Ma old. One is a sequence of rocks in eastern South Africa known as the Swaziland Supergroup. This group forms the hills and valleys of the Barberton Mountain Land in and near the kingdom of Swaziland. The other, a thick pile of volcanic and sedimentary rocks named the Pilbara Supergroup, crops out as ridges, buttes, and broad domical hills west of the Great Sandy Desert near the northwestern corner of Western Australia. Both terrains have yielded convincing fossils and carbon

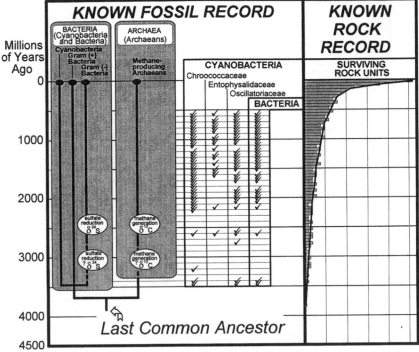

Figure 6.7. Comparison of the distributions in time of known Precambrian
microbial fossils and rock units that have survived to the present. The
checkmarks in the central columns show the spread in time of thousands of
fossil occurrences known from Precambrian rocks. Each check signifies that
microbes of the kind listed have been found in the 50-Ma-long interval
indicated.

isotopes indicating photosynthesis, but the less pressure-cooked
Australian rocks contain more abundant, better preserved records of life.

The best preserved and oldest of the fossils are minute, threadlike
microbes (figure 6.8). Eleven kinds of these microbes have been found
entombed in a dark gray, carbon-rich rock unit known as the Apex
chert, which weathers out as ledges and large angular blocks in low
hills near the one-horse town of Marble Bar. (This tiny town is about
150 km southeast of Port Hedland, the coastal settlement at the northwestern edge of Australia that serves as the hub of the Pilbara region.)
The fossil-bearing bed was laid down along the margin of a narrow
seaway flanked by towering volcanoes that episodically blanketed the

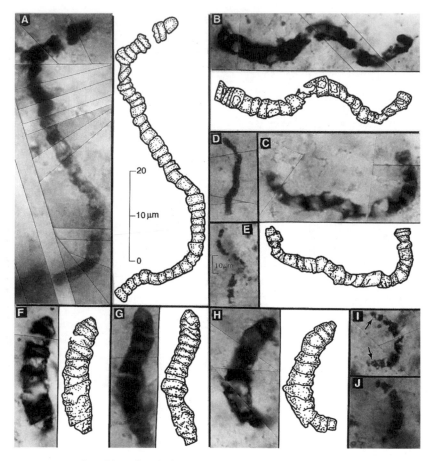

Figure 6.8. The oldest fossils known, cyanobacterium-like filamentous microbes shown in thin slices of the nearly 3,500-Ma–old Apex chert of northwestern Western Australia. The magnification of (D, E, I, and J) is shown by the scale in (E); the magnification of all other parts is shown by the scale in (A). (A–E) *Primaevifilum amoenum;* (F–J) *Primaevifilum conicoterminatum;* arrows in (I) point to conical end cells.

seafloor with massive sheets of lava. Sandwiched between two of these volcanic flows, each precisely dated, the fossiliferous horizon is 3,465 ± 5 Ma old, nearly three-quarters the age of the Earth.

Even though the Pilbara rocks comprise the least pressure-cooked exceptionally ancient geologic sequence known to science, the Apex fossils are scrappy, hard to find, and difficult to study. They are abundant, but they are also mangled, shredded, and charred. Tiny bits and pieces are common, but generally nondescript; short two- or three-celled fragments are rare and easy to overlook; many-celled specimens are few and far between; and truly well preserved fossils are nonexistent. But given their immense age—and their fragile makeup and their minute size—it is remarkable that they have survived at all. Like the preserved growth rings of a fossil log, the Apex microbes are petrified, composed of coaly remnants of their original cell walls, the living juices leached away long ago and the cells filled in with mineral. Almost nothing is left to establish their identity except the size, shape, and boxcar-like arrangement of their simple filament-forming cells. But these clues, gathered from 2,000 cells in hundreds of examples of 11 kinds of fossils, reveal how the cells divided and multiplied. And they offer good reasons for identifying these microbes as members of the bacterial domain, most likely cyanobacteria and various kinds of noncyanobacterial bacteria.

Several of the Apex species seem almost indistinguishable from living cyanobacteria of the taxonomic family **Oscillatoriaceae,** microbes that are common today in oceans, lakes, and forest soils worldwide. The existence very early in the Precambrian of this particular group of cyanobacteria fits with rRNA trees, which show them to be one of the most primitive kinds of cyanobacteria living today. It also fits with the abundant presence of members of this group in the more recent Precambrian fossil record (figure 6.7) where, like the Apex fossils, their petrified threadlike filaments are the chief components of biologically diverse shallow-water microbial communities. But other fossils of the Apex assemblage more closely resemble noncyanobacterial bacteria. Thus, the signature of photosynthesis recorded in the Apex organic matter may reflect the presence of both kinds—cyanobacteria and their close relatives, photosynthetic bacteria.

Cemented to rocks and boulders on the seaway floor by a thick layer of sticky mucilage, the Apex community was protected by overlying waters from lethal **UV light** that, in the absence of an oxygen-rich (and, hence, of an **ozone**-rich) atmosphere, flooded the surface of the early Earth. Thus cosseted, it was a biologically diverse microbial menagerie. This oldest known fossiliferous deposit gives but a scanty glimpse of the ancient living world; but the preservation, even in a single fossil horizon, of such a varied suite of microbes is a clear indicator that a profusion of microorganisms already existed at this very early stage in Earth's history. By 3,500 Ma ago, diverse microbes, many like those living today, were already thriving, not only in what is now Western Australia but in habitable settings across the globe.

Strengths and Weaknesses of Isotopes and Fossils

Direct evidence from the geologic record—both of carbon isotopic signatures and cellularly preserved fossils—establishes beyond doubt the
great antiquity of life. Linked to the present by a continuous chain of
paleobiologic evidence (figure 6.7), life on Earth dates from at least 3,500
Ma ago. Only a billion or so years after Earth's formation its surface
swarmed with living systems. This microbial zoo evidently included
cyanobacteria—microorganisms notable as the biochemically most
advanced; the morphologically, physiologically, and ecologically most varied; and, in cell size, by far the largest members of the bacterial domain.
Clearly, life evolved far and fast very early in the history of the planet.

But direct evidence from the rock record can take us only part of
the way toward unraveling the mysteries of life's beginnings. Truly
ancient rocks are few and they have mostly been subjected to evidence-
destroying pressure-cooking. Moreover, the known rock record extends back no further than 3,800 Ma ago, leaving a huge gap of un-
documented Earth history because the planet formed some 700 million
years earlier. The oldest fossils known are varied and highly evolved,
much too advanced to shed light on what the earliest life forms were
like. And, in any case, the oldest fossils represent no more than the
oldest *detected* cellular life, not its actual earliest presence. Evidence
from the rock record sets a minimum age for life's existence, but it
does not and cannot provide a full answer to the *what* and *when* of
life's beginnings.

WHEN WAS EARTH READY FOR LIFE?

The paleobiologic record tells us that life arose earlier than 3,500 Ma
ago, but just how much earlier is an open question. Graphite in the
3,800-Ma–old Isua rocks of Greenland hints that life may have existed
then but is too altered by heat and pressure to prove that life was
present. And rocks are totally unknown from the earlier 700 million
years of Earth's existence—there is no surviving earthly evidence to
reckon with.

What happened during the embryonic stages of Earth history, the
hundreds of millions of years missing from the geologic record? The Moon
holds this answer. Too small to hold an atmosphere or ocean, the Moon
does not have huge continents or massive mountain chains. There is
no wind, it never rains, and rocks don't weather as they do on Earth.

IMPACT FRUSTRATION OF THE ORIGIN OF LIFE

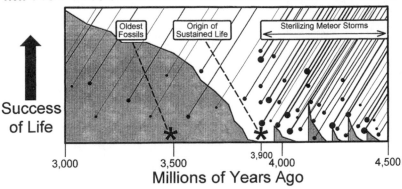

Figure 6.9. The common ancestor of life today could have originated only after the last great planet-sterilizing meteoritic impact.

Though churned by **meteorites** over the ages, the otherwise pristine lunar surface retains a record of its youth, a catastrophic epoch indelibly inscribed in the scarred and cratered moonscape.

During the segment missing from our planet's geology, the Moon, the Earth, and other bodies of the Solar System were in the final stages of formation, sweeping up huge chunks of rocky debris encountered in their orbits. They were bombarded and blasted by in-falling meteors. And because the Earth is so much larger than the Moon, the barrage here was much greater. The heat of the bombarding fragments melted out the Earth's iron core, which settled to the center of the forming planet. Lighter "rockbergs" floated on its seething molten surface, beginning to congeal into a solid rocky rind about 4,300 Ma ago. So intense was the bombardment and so enormous the in-falling meteors that the Earth's oceans repeatedly boiled away. Even a medium-sized impactor 5 km across would have turned the oceans to steam, shrouding the planet in a dense, lightning-charged, pitch-black cloud bank that would have lasted thousands of years before it rained out to form the oceans again. Had any life gained a foothold in this hellish setting, it would have been wiped out as the entire planet was sterilized over and over again (figure 6.9).

Catastrophic collisions have happened throughout the history of our planet (and are sure to happen again). Sixty-five million years ago a rocky mass 10 km wide smashed into the Yucatán Peninsula, bringing tidal waves, wildfires, choking smoke, and flaming ashes that decimated the living world and helped wipe out the dinosaurs. In 1908, a much smaller body, 70 meters across, blew up over Tunguska, Siberia. The 20-megaton blast ignited the clothes of a man some 100 km away, annihilated herds of reindeer, and flattened thousands of acres of dense-packed forest. Every 100,000 years or so, Earth is hit by a kilometer-sized meteorite that throws enough debris into the atmosphere to block out the sun, plunging the world into a years-long night, nature's equivalent of nuclear winter.

But even such terrifying events pale to insignificance in comparison with the deadly firestorm that rained onto the primal Earth for hundreds of millions of years. The sterilizing devastation kept on until about 3,900 Ma ago when the last of the huge orbiting chunks was swept away. Living systems may have originated and been killed off many times during earlier planetary history—there is no way to tell. But life as we know it could come into being only after 3,900 Ma ago, and a scant 400 Ma later, it was flourishing and widespread. How did life advance so far, so fast?

So Far, So Fast, So Early?

We have no fossils that testify to evolution's rapid progress from 3,900 to 3,500 Ma ago. But we have other cards up our sleeve. Perhaps the most promising is the fossil record of the most recent 400 million years, an especially well-studied chapter in life's history. This record should give us a good reading of how much evolution can happen in a relatively short period of geologic time.

What was life's lot during the most recent 400 million years? First, primitive land plants appeared, tiny twigs less than a centimeter high. From these paltry beginnings, plants soon evolved, producing lush lowland vegetation, the giant scale-trees of the Coal Swamp Floras, dense highland forests, and, eventually, rich grasslands and all the trees, shrubs, and flowering plants of the modern world. While this was going on, *all* animal life on land evolved as well—first, amphibians such as salamanders and frogs; then a multitude of reptiles, most notably the dinosaurs and their kin, including the feathered dinosaurians we call birds; and, finally, warm-blooded hairy mammals, including primates such as ourselves. Clearly, an enormous amount of evolution can be squeezed into a scant 400 million years.

But the impressive feats of recent geologic time may not be a fair test of the speed of early evolutionary advance. By 400 Ma ago, life had long been in place; yet if organisms existed 3,900 Ma ago, they were

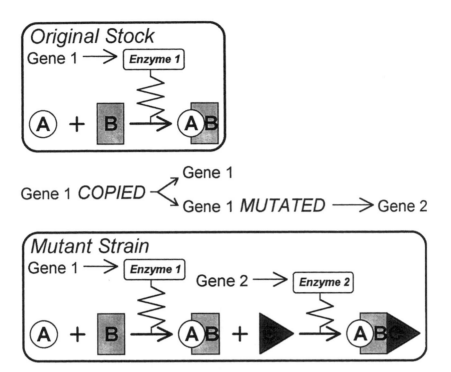

Figure 6.10. Gene copying speeded early evolution by adding new biochemical functions without affecting those already in place. In the upper panel, Gene 1 contains the genetic information responsible for formation of Enzyme 1, a matchmaker that links molecules A and B to make compound AB. Gene 1 is then copied, and one of the two identical versions is mutated to Gene 2, which, as shown in the lower panel, is responsible for formation of Enzyme 2, a new matchmaker that links AB to molecule C to form compound ABC. By mutation of a copied gene, a new function has evolved while the original function has been retained.

primitive and simple. How could fledgling life have evolved so far and fast? The mechanisms of early evolution appear to hold the key. The hallmark of the evolutionary process is its conservatism and economy, its incredibly successful strategy of building, bit by bit, on what already exists. A prize way to accomplish this—**gene copying**—came into being early in life's history. As long as one copy of a gene functions, the extras can be mutated, sometimes to genes that work even better or do new tricks (figure 6.10). And because new genes don't have to be made from scratch, this is a surefire mechanism for speedy evolution. The advent of the various mechanisms of physiology and metabolism followed a similar pattern. New ways were invented early to meet life's demand for CHON and cellular energy; but instead of concocting totally novel schemes to do these jobs, evolution took shortcuts by recasting and reusing bits and pieces of similar processes already in place. Gene xeroxing and biochemical remodeling—both examples of the conservatism and economy of the evolutionary process—played major roles in speeding life to its early advance.

THE *WHAT* AND *WHEN* OF LIFE'S BEGINNINGS

Although we have made enormous progress toward unraveling the mysteries associated with the origin of life, hard evidence of the *what* of life's beginnings remains elusive. Neither indirect evidence from modern organisms (the Universal Tree) nor direct evidence from the rock record (the findings of paleobiology) can take us more than part way toward defining the makeup of the earliest life forms. If a firm answer is to be found, it most likely will be detected in the molecular biology of living microbes. And we have just begun to unearth this treasure-trove of promising clues.

The *when* of life's beginnings can be answered more fully, though not yet precisely. Life's origin dates certainly from before 3,500 Ma ago (as shown by paleobiology), probably from after 3,900 Ma ago (and the last of the planet-sterilizing impact events), and perhaps from as early as 3,800 Ma ago (as hinted by the Isua graphite). And this range is likely to be narrowed. It would be a great surprise if fossils more ancient than the oldest now known are not discovered, and perhaps soon—an allout search of the ancient rock record is only now beginning, its potential barely tapped.

FURTHER READING

Groves, D. I., J. S. R. Dunlop, and R. Buick. 1981. An early habitat of life. *Scientific American* 245:64–73.

Islas, S., A. Becerra, J. I. Leguina, and A. Lazcano. 1998. Early metabolic evolution: Insights from comparative cellular genomics. In *Exobiology: Matter, energy, and information in the origin and evolution of life in the Universe*, ed. J. Chela-Flores and F. Raulin, 167–74. Amsterdam: Kluwer.

Iwabe, N., K. Kuma, M. Hasegawa, S. Osawa, and T. Miyata. 1989. Evolutionary relationships of archaebacteria, eubacteria, and eukaryotes inferred from phylogenetic trees of duplicated genes. *Proceedings of the National Academy of Sciences USA* 86:9355–59.

Lazcano, A. 1995. Cellular evolution during the early Archean: What happened between the progenote and the cenancestor? *Microbiologia SEM* 11:1–13.

Lazcano, A. and P. Forterre, eds. 1999. Special Issue: The Last Common Ancestor and beyond. *Journal of Molecular Evolution* 49:411–540.

Lazcano, A., G. E. Fox, and J. Oró. 1992. Life before DNA: The origin and evolution of early Archean cells. In *The Evolution of metabolic function,* ed. R. P. Mortlock, 237–95. London: CRC Press.

Olsen, G. J., and C. R. Woese. 1993. Ribosomal RNA: A key to phylogeny. *FASEB Journal* 7:113–23.

_____. 1997. Archaeal genomics: An overview. *Cell* 89:991–94.

Penny, D., and A. Poole. 1999. The nature of the last universal common ancestor. *Current Opinion in Genetics & Development* 9:672–77.

Schopf, J. W. 1993. Microfossils of the Early Archean Apex chert: New evidence of the antiquity of life. *Science* 260:640–46.

_____. 1999. *Cradle of Life, The Discovery of Earth's Earliest Fossils.* Princeton, New Jersey: Princeton University Press.

Sleep, N. H., K. J. Zahnle, J. F. Kasting, and H. J. Morowitz. 1989. Annihilation of ecosystems by large asteroid impacts on the early Earth. *Nature* 342:139–42.

Stetter, K. O. 1996. Hyperthermophilic procaryotes. *FEMS Microbiological Review* 18:149–58.

_____. 1996. Hyperthermophiles in the history of life. In *Evolution of hydrothermal ecosystems on Earth (and Mars?),* ed. G. R. Bock and J. A. Goode, 1–10. New York: Wiley.

Woese, C. R. 1987. Bacterial evolution. *Microbiological Reviews* 51:221–71.

Woese, C. R., O. Kandler, and M. L. Whellis. 1990. Towards a natural system of organisms: Proposal for the domains Archaea, Bacteria, and Eucarya. *Proceedings of the National Academy of Sciences USA* 87:4576–79.

Contributors

SHERWOOD CHANG, SETI INSTITUTE, MOUNTAIN VIEW, CALIFORNIA

Sherwood Chang, an international leader in cosmochemistry and studies of the origin of life, has been a friend and colleague of Alan Schwartz (coauthor of chapter 2) for many years, an association dating from the late 1960s when they shared laboratory facilities at NASA's Ames Research Center south of San Francisco. Following completion of his undergraduate studies at Harvard (A. B. in chemistry, 1962), receipt of his doctorate from the University of Wisconsin (in organic chemistry, 1966), and postdoctoral research at Stanford University, Dr. Chang joined the staff at NASA–Ames in 1967, first as a Staff Scientist and Principal Investigator, then as Chief of the Exobiology Branch, a position he filled with great distinction from 1985 to 1998. Currently, he is formally affiliated with the SETI Institute in northern California, a not-for-profit research organization focusing on the search for extraterrestrial life. For his contributions in cosmochemistry, geochemistry, and lunar and planetary science, Dr. Chang was awarded NASA's Exceptional Scientific Achievement Medal, two NASA Special Achievement Awards, the H. Julian Allen Award, and a prestigious Dryden Fellowship. From 1986 through 1993, he served as First Vice President of the International Society for the Study of the Origin of Life, the worldwide membership of which elected him an *ISSOL* Fellow in 1999.

JAMES P. FERRIS, RENSSELAER POLYTECHNIC INSTITUTE,
TROY, NEW YORK

James P. Ferris is the 1996 recipient of *ISSOL*'s A. I. Oparin Medal, awarded for his masterful studies of how small building-block molecules (monomers) in the primordial environment could have linked to form long chainlike aggregates (polymers), compounds such as RNA and proteins on which life depends. He is currently a Research Professor and Director of the New York Center for Studies on the Origin of Life (a NASA NSCORT) at Rensselaer Polytechnic Institute, whose faculty he joined in 1967. Trained in chemistry at the University of Pennsylvania (B. S., 1954) and Indiana University (Ph.D., 1958), he was a member of the Department of Chemistry at Florida State University and a Research Scientist at the Salk Institute for Biological Studies in southern California. He has been a Visiting Research Scientist at the State University of New York, Albany; the Marine Biology Laboratories, Woods Hole, Massachusetts; the Salk Institute for Biological Studies; the NASA Ames Research Center, California; and the Eidgenössische Technische Hochschule, Laboratory of Organic Chemistry, Zürich, Switzerland. A past president of *ISSOL*, he was a long-term (1982–1999) editor of the Society's journal, *Origins of Life and Evolution of the Biosphere*, helping to achieve its current international preeminence.

ANTONIO LAZCANO, AUTONOMOUS NATIONAL UNIVERSITY, MEXICO CITY

Antonio Lazcano, a close colleague and long-time friend of Stanley Miller (coauthor of chapter 3), is among the best known and most respected of all modern Mexican biologists. Trained both as an undergraduate and graduate student at Mexico's Autonomous National University in Mexico City, he is a recipient of that university's most prized accolade—the title of Distinguished Professor. Recipient of the National University's Gold Medal of Biological Research, he has also received an impressive series of other prestigious awards for his contributions to science, scientific journalism, and teaching. An academic committed to public education, Professor Lazcano is the author of several books including *The Miraculous Bacteria*, a tour de force on microbial evolution; the *Spark of Life*, a layperson-level exploration of how life on Earth began; and a national bestseller (more than 350,000 copies), *The Origin of Life*. An internationally recognized scholar, he has been a Professor-in Residence or Visiting Scientist at universities and research institutes in France, Spain, Cuba, Switzerland, Russia, and the United States. Currently, the First Vice President of the International Society for the Study of the Origin of Life, Professor Lazcano focuses his research on the deepest branches of the Universal Tree of Life and the origin and earliest evolution of the energy-yielding metabolic pathways of living systems.

STANLEY L. MILLER, UNIVERSITY OF CALIFORNIA, SAN DIEGO

Stanley L. Miller, *ISSOL*'s 1983 A. I. Oparin medalist, is legendary for his fundamental contributions to understanding life's beginnings—contributions that are now recounted in textbooks worldwide. An early pathfinder in origin-of-life research, he was the first to show, in experiments plausibly simulating pre-life Earth conditions, that amino acids and other building blocks of life could have been formed in the absence of living systems. A native of Oakland, California, Professor Miller received his undergraduate training at the University of California, Berkeley, and his doctorate in chemistry (1954) from the University of Chicago. Following completion of an F. B. Jewett Fellowship at the California Institute of Technology, and after serving as a faculty member in the College of

Physicians and Surgeons at Columbia University in New York City, he joined the University of California, San Diego, where he has been Professor of Chemistry since 1958. He was the first Director of its NASA-supported NSCORT, a Specialized Center for Research and Training in origin-of-life studies. A past president of *ISSOL*, member of the National Academy of Sciences, and an Honorary Councilor of the Higher Council for Scientific Research of Spain, Professor Miller has been recognized by the *Los Angeles Times Magazine* as among the most outstanding scientists of southern California in the 20th century.

LESLIE E. ORGEL, THE SALK INSTITUTE, LA JOLLA, CALIFORNIA

Leslie E. Orgel, *ISSOL*'s 1993 H. C. Urey medalist, is an innovative visionary and chemist extraordinaire whose experimental studies have laid the groundwork for understanding how information-containing molecules may have emerged and evolved on the primitive Earth. Born in London, Dr. Orgel was educated in chemistry at Oxford University (B. A., 1949; Ph.D., 1951), where he was a Fellow of Magdalen College from 1950 to 1954. His early career focused on theoretical inorganic chemistry (at Oxford; the California Institute of Technology; the University of Chicago; and the University of Cambridge, England), research for which he was awarded the Harrison Prize and elected to the Royal Society. His interests then turned to biological problems. In 1964, he was appointed to his present position of Senior Fellow and Research Professor at the Salk Institute for Biological Studies at La Jolla, California, where he directs the Chemical Evolution Laboratory. A member of the National Academy of Sciences and recipient of a Guggenheim Fellowship, Dr. Orgel has been a Visiting Professor at the Hebrew University, Jerusalem; the Massachusetts Institute of Technology; and Yale University. Respected worldwide for his contributions to science, he is also known as a gifted colleague/mentor of outstandingly successful researchers (such as *ISSOL* medalists Alan Schwartz and James Ferris).

JOHN ORÓ, UNIVERSITY OF HOUSTON, TEXAS

John Oró is the 1986 recipient of *ISSOL*'s A. I. Oparin Medal, awarded for his breakthrough experimental studies of prebiotic chemistry, especially of how the components of gene-carrying nucleic acids may have first formed on the primitive Earth. He also was among the first to suggest a major role for comets as a source of pre-life organic compounds, a view now widely accepted, and is well known for his pioneering studies of the organic constituents of meteorites and ancient rocks. Founder and professor emeritus of the Department of Biochemical and Biophysical Sciences at the University of Houston, Texas, he is the recipient of doctorates from the University of Houston and the Universities of Lleida and Granada in Spain. Professor Oró was born in Catalonia, in the Spanish town of Lleida near Barcelona—a region he represented as an elected senator to the Catalonian Congress. He now serves as president of Fundació Joan Oró, a not-for-profit foundation actively involved in public science education. A past president of *ISSOL* and a scientist who made important contributions to the success of NASA's Apollo and Viking programs during the late 1960s and 1970s, he is the author of more than 30 books and 300 scientific articles and has organized and convened some 30 scientific symposia and international scientific meetings.

J. WILLIAM SCHOPF, UNIVERSITY OF CALIFORNIA, LOS ANGELES

J. William Schopf is the Director of UCLA's Center for the Study of Evolution and the Origin of Life and a member of the Department of Earth and Space Sciences. Currently *ISSOL*'s president, he was awarded the Society's 1989 A. I. Oparin Medal for his contributions to understanding the early (Precambrian) history of life. Born in Urbana, Illinois, he received his undergraduate training in geology at Oberlin College, Ohio, and obtained his doctorate in biology from Harvard in 1968. Author of *Cradle of Life,* winner of the 2000 Phi Beta Kappa Science Award, and editor of two national prize-winning monographs on early evolution, Professor Schopf has received all three UCLA campus-wide awards for Teaching, Research, and Academic Excellence and has been recognized by *Los Angeles Times Magazine* as among southern California's most outstanding scientists of the 20th century. He has been honored internationally as a Senior Humboldt Fellow in Germany, a Foreign Member of the Linnean Society of London, and an honorary member of the faculties of Yunnan Teachers University, China, and the Russian Academy of Science's A. N. Bakh Institute of Biochemistry. A member of the National Academy of Sciences, the American Philosophical Society, and the American Academy of Arts and Sciences, he is the recipient of two Guggenheim Fellowships and of medals awarded by the National Science Board–National Science Foundation and by the National Academy of Sciences.

ALAN W. SCHWARTZ, UNIVERSITY OF
NIJMEGEN, TOERNOOIVELD, THE NETHERLANDS

Alan W. Schwartz is the 1999 recipient of *ISSOL*'s H. C. Urey Medal, awarded for his exemplary interdisciplinary studies of some of the most vexing questions about how life began: What role was played by organic matter brought to Earth from other bodies? What was the source of phosphate on the primordial planet? What was the nature of the RNA World, the primitive rootstock of today's DNA-based life, and how did it get started? A native of New York City, he was educated in chemistry and biochemistry at New York University (B. A., 1957) and Florida State University (Ph.D., 1965). Before joining the faculty at the University of Nijmegen, The Netherlands, where he has been Professor of Exobiology since 1968, he held positions at Los Alamos Scientific Laboratory, New Mexico, and at NASA's Ames Research Center, California. Chair of the Council of Europe's Research Group on Cosmic Chemistry, Chemical Evolution, and Exobiology, and advisor to the European Space Agency Exobiology Team, Professor Schwartz has been a Visiting Scientist both at the Salk Institute for Biological Studies in southern California and the University of Tsukuba, Japan. From 1972 to 1999, he was editor of the international journal *BioSystems,* a position he has now assumed for *Origins of Life and Evolution of the Biosphere,* the official journal of the International Society for the Study of the Origin of Life.

Glossary

Acetaldehyde An organic aldehyde, CH_3CHO.

Acetic acid An organic acid, CH_3COOH.

Acetonitrile An organic cyanide, CH_3CN.

Acetylene An organic compound, the hydrocarbon C_2H_2.

Achiral See *Nonchiral*.

Activating group Any of various chemical groups whose attachment to molecules primes those molecules for chemical reaction (e.g., by promoting formation of linking bonds between two monomers).

Adaptive Radiation The evolution of a species into a group of species adapted to different ecological niches.

Adenine (A) A purine ($C_5H_5N_5$), a component of ATP, and one of the four nitrogenous bases present in the nucleic acids DNA and RNA. (For chemical structure, see figures 3.5 or 4.1a.)

Adenosine triphosphate (ATP) An organic compound ($C_{10}H_{16}N_5O_{13}P_3$) present in all cells that can release energy (derived from sunlight in photoautotrophs, chemical reactions in chemoautotrophs, or food in heterotrophs), stored in its phosphate bonds, to power cellular processes.

Aerobe Any of various organisms capable of using molecular oxygen (O_2) in aerobic respiration, either as a requirement for living processes (as in obligate aerobes) or as an energy-yielding process to supplement fermentation (as in facultative aerobes).

Aerobic respiration The oxygen-consuming, energy-yielding process carried out by almost all eukaryotes and diverse prokaryotes.

Alanine A chemically neutral amino acid, $CH_3CH(NH_2)COOH$, one of the 20 amino acids commonly present in proteins of living systems.

Alcohol Any of various hydroxyl group–containing organic compounds having the general formula ROH in which R is a univalent hydrocarbon radical.

Aldehyde Any of various CHO-containing organic compounds having the general formula RCOH in which R is a univalent hydrocarbon radical.

ALH84001 The 1.9-kg meteorite claimed to contain evidence of past Martian microbial life; named for where and when it was found (Allan Hills ice field, Antarctica, in 1984) together with its assigned sample number (001).

Aliphatic Any of various open-chain (noncyclic) hydrocarbons.

Alpha-amino-n-butyric acid (*α-amino-n-butyric acid*) An amino acid, $CH_3CH_2CH(NH_2)COOH$, not commonly present in proteins of living systems.

Alpha-aminoisobutyric acid (*α-aminoisobutyric acid*) An amino acid, $(CH_3)_2C(NH_2)COOH$, not commonly present in proteins of living systems.

Alpha-helix (*α-helix*) A simple spiral arrangement of component parts; characteristic of the helical structure of portions of some proteins and the double-stranded DNA of chromosomes.

Amide An organic compound resulting from replacement of one or more atoms of hydrogen in ammonia by univalent acid radicals.

Amide bond See *Peptide bond*.

Amine Any of various organic compounds derived from ammonia by replacement of hydrogen by one or more univalent hydrocarbon radicals.

Amination The addition of an amino group, $-NH_2$, to a compound.

Amino acid A small molecule (monomer) having an amino group ($-NH_2$) at one end and a carboxylic acid group ($-COOH$) at the other; amino acids can be linked to form a protein having the general formula $RCH(NH_2)COOH$ in which R is a univalent hydrocarbon radical.

Amino acid amide An amino acid–related compound characterized by replacement of the hydroxyl ($-OH$) of the amino acid carboxylic acid group ($-COOH$) with the amino group ($-NH_2$), resulting in the general formula $RCH(NH_2)CONH_2$ in which R is a univalent hydrocarbon radical.

Amino acid nitrile An organic cyanide compound having the general formula $RCH(NH_2)CN$ in which R is a univalent hydrocarbon radical.

Amino amide See Amino acid amide.

Amino group The three-atom chemical group $-NH_2$.

Aminomaleonitrile An organic compound, $NC(NH_2)CHCN$. (For chemical structure, see chapter 3.)

Amino nitrile See Amino acid nitrile.

Aminoimidazole carbonitrile An organic compound, $H_4C_4N_4$. (For chemical structure, see chapter 3.)

Ammonia The chemical compound NH_3.

Ammonium cyanate An organic compound, $NCONH_4$.

Anabolism Metabolism resulting in buildup (biosynthesis) of organic compounds.

Anaerobe Any of various organisms, almost all prokaryotes, that can live in the absence of molecular oxygen (O_2).

Anoxygenic photosynthesis The non–oxygen-producing photoautotrophy of photosynthetic bacteria.

Apex chert A fossiliferous horizon of the Apex Basalt, a 3,465 ± 5 Ma-old geologic unit of Western Australia.

Archaea (archaeal domain) Together with Bacteria and Eucarya, one of three superkingdom-like primary branches of the Tree of Life.

Archaean microorganism Any of diverse microbes of the Archaea, a domain of non-nucleated microbes that includes many kinds of extremophiles (microorganisms able to thrive in exceedingly acidic high temperature settings) and methanogens (microbes that give off methane gas as a product of metabolism).

Arginine An amino acid, $C_6H_{14}N_4O_2$, one of the 20 amino acids commonly present in proteins of living systems.

Aromatic Pertaining to any of various hydrocarbons characterized by the presence of at least one benzene ring, C_6H_6.

Artificial life Pertaining to "lifelike" machines (e.g., robots and computers) and similarly nonbiologic constructions that exhibit biologic-like properties.

Aspartic acid An acidic amino acid, $HOOCCH_2CH(NH_2)COOH$, one of the 20 amino acids commonly present in proteins of living systems.

Asteroid Any of thousands of small planetlike bodies orbiting the Sun in the region of space between Mars and Jupiter and having diameters from a fraction of a kilometer to about 1,000 km.

Asteroidal belt The zone of space between the orbits of Mars and Jupiter.

Astronomical unit (AU) The mean distance between the Earth and the Sun, about 150 million km (93 million miles).

ATP See *Adenosine triphosphate.*

Autocatalytic Pertaining to a chemical reaction that is catalyzed by the products it itself forms.

Autotrophy A metabolic process of plant and plantlike organisms (photoautotrophy) and diverse nonphotosynthetic bacteria and archaeans (chemoautotrophy) in which carbon dioxide serves as the principal source of cellular carbon.

Bacteria (bacterial domain) Together with Archaea and Eucarya, one of three superkingdom-like primary branches of the Tree of Life.

Bacterial microorganism Any of diverse microbes of the Bacterial domain composed of non-nucleated small-celled life forms that includes all of the common bacteria as well as cyanobacteria.

Bacterium Any of diverse prokaryotes, including cyanobacteria, of the Bacterial domain.

Bar A unit of pressure equal to one dyne per square centimeter.

Barbituric acid An organic acid, $C_4H_4N_2O_3$.

Base sequence The genetic information–encoding aperiodic arrangement of nitrogenous bases (the purines, adenine and guanine; and the pyrimidines, cytosine and thymine or uracil) in DNA or RNA.

Benzene An aromatic hydrocarbon, C_6H_6.

Beta-alanine (β-alanine) An amino acid, $NH_2CH_2CH_2COOH$, not commonly present in proteins of living systems.

Beta-sheet (β-sheet) A flat sheetlike arrangement of component parts, such as that characteristic of portions of some proteins.

Bicarbonate An inorganic chemical, HCO_3^-, formed when carbon dioxide dissolves in water.

Big Bang The explosive birth of the Universe in a very hot dense state, about 13 to 15 billion years ago, followed by the expansion of space.

Biochemical Any of a large number of chemical compounds made by living systems, commonly composed of a mixture of the biogenic elements.

Biochemistry Chemistry that deals with the chemical compounds and processes that occur in organisms.

Biogenic elements The chemical elements generally regarded as necessary for life: carbon, hydrogen, oxygen, nitrogen, sulfur, and phosphorus.

Biopolymer Any of various polymeric chemical compounds made by living systems such as proteins (polymers of amino acids) and carbohydrates (polymers of sugars).

Biosynthesis The process of manufacture of organic compounds by biologic systems.

Biosynthetic pathway Any of diverse enzyme-mediated multistep processes by which organic compounds are formed in living systems.

Biotin An organic compound, $C_{10}H_{16}N_2O_3S$, a growth factor present in very minute amounts in every living cell.

Bond energy Chemical energy that holds together the atoms of a molecule.

Butadiene An organic compound, the hydrocarbon C_4H_6.

Butlerov (Butlerow) synthesis Known also as the formose reaction, a series of chemical reactions, first described in 1861 by A. M. Butlerov, by which treatment of formaldehyde with a strong alkaline catalyst (e.g., NaOH) leads to the formation of sugars.

Calcium carbonate An inorganic compound, $CaCO_3$, commonly present as the mineral calcite that makes up limestones.

Calcium hydroxide The inorganic base $Ca(OH)_2$.

Cambrian Period The earliest geologic period of the Phanerozoic Eon of Earth's history, extending from 543 to 495 Ma ago.

Carbide Any of various carbon–metal compounds.

Carbohydrate A carbon-, hydrogen-, and oxygen-containing organic polymeric compound composed of sugar monomers.

Carbon dioxide A colorless odorless gas, CO_2, a component of Earth's atmosphere.

Carbon fixation The metabolic incorporation of carbon into an organic compound.

Carbon monoxide A colorless odorless gas, CO, toxic to aerobic organisms.

Carbon-14 (^{14}C) A radioactive isotope of carbon produced in the upper atmosphere and present in living plants and animals; used in carbon-14 dating because, by giving off a beta ray, it decays to nitrogen (^{14}N) with a half-life of about 5,730 years.

Carbonaceous Containing or composed of coal-like organic matter (kerogen), whether nonbiologic (as in carbonaceous chondrites) or biologic (as in coal and organic-walled fossils).

Carbonaceous chondrite A relatively rare type of stony meteorite, so named because of the presence of tiny glassy balls (chondrules) and as much as 5% of the element carbon, largely in the form of "coaly" organic matter (kerogen) formed nonbiologically.

Carbonaceous meteorite A meteorite of the carbonaceous chondrite group.

Carbonate Any of various minerals containing the chemical group $CO_3{}^{2-}$, such as calcite ($CaCO_3$) or dolomite ($CaMg\ [CO_3]_2$); or a rock consisting chiefly of such minerals, such as limestone or dolostone.

Carbonyl diimidazole An organic compound, the chemical condensing agent shown in figure 4.16a.

Carboxyl (carboxylic acid) group The four-atom chemical group –COOH characteristic of carboxylic organic acids.

Catabolism Metabolism resulting in breakdown of organic compounds and release of energy.

Catalysis Modification, especially an increase of rate, of a chemical reaction induced by a substance such as an enzyme or ribozyme that is unchanged chemically at the end of the reaction.

Centrifugal force The force that tends to impel an object outward from the center of rotation.

Chemical bond A linkage between two atoms of a molecule or between atoms of neighboring molecules, often by shared electrons.

Chemoautotrophy An autotrophic metabolism in which energy is generated by oxidation of an inorganic compound.

Chemolithotrophy A chemically driven metabolic process commonly powered by coupled oxidation–reduction reactions of inorganic compounds.

Chert A type of rock composed of microcrystalline quartz (SiO_2).

Chimera An individual consisting of parts of diverse genetic constitution (for humans, with reference especially to mitochondria, organelles where aerobic respiration occurs in cells, derived initially in early eukaryotes from free-living purple bacteria).

Chirality Pertaining to the "handedness" of stereoisomers of compounds present either in the D-configuration (so named because pure solutions of such compounds rotate plane-polarized light in the *dextro*, rightward direction) or the L-configuration (pure solutions of which rotate plane-polarized light to the *levo*, leftward direction).

Chlorophyll Any of several structurally similar light-absorbing pigments that play a central role in the photosynthesis of cyanobacteria, algae, and higher plants.

Chondrule A particular kind of small mineralic glass spheroid present in chondritic meteorites.

CHONSP Abbreviation for the biogenic elements: carbon, hydrogen, oxygen, nitrogen, sulfur, and phosphorus.

Circumstellar disk A nebula of gas and dust orbiting a star, typically having radial sizes comparable to that of the Solar System.

Citric acid cycle The electron transport cycle of aerobic respiration.

CITROENS Abbreviation of a definition of life coined by L. E. Orgel: Complex Information-Transforming Reproducing Objects that Evolve by Natural Selection.

Clay mineral Any of a group of hydrous silicates of aluminum chiefly formed in weathering processes and present especially in clays and shales.

Clay World A stage in prebiotic, precellular, evolution hypothesized by G. Cairns-Smith during which life originated as a self-replicating inorganic clay that evolved into organic-based living systems.

Cloud core Any of various regions in the interstellar medium containing relatively high concentrations of dust and gas (of the order 10^5 to 10^6 molecules/cm^3) that serve as spawning grounds for the origin of stars.

CNO cycle A linked cycle of nucleosynthetic reactions by which nitrogen and other nuclides are generated catalytically.

Coacervate A viscous aggregate of colloidal microscopic droplets held together by electrostatic forces.

Coenzyme A protein that forms the active portion of an enzyme system in combination with one or more other substrate-specific enzymes.

Comet A celestial body that consists of a core usually surrounded by a bright nucleus, a dirty snowball-like object composed of aggregated very low-temperature interstellar matter.

Comet, long period Cometary bodies whose passage in the vicinity of Earth occurs over relatively long time periods; thought to be derived from the Oort cloud.

Comet, short period Cometary bodies whose passage in the vicinity of Earth occurs over relatively short time periods; thought to be derived from the Kuiper belt.

Condensing agent Chemicals such as cyanogen, cyanamide, and cyanoacetylene that facilitate polymerization of monomers.

Cosmochemistry A branch of chemistry that focuses on the composition of the cosmos.

Cosmology The study of the overall structure and evolution of the Universe.

Covalent bond A bond formed between atoms by the sharing of electrons.

Cretaceous Period The youngest of the three geologic periods of the Mesozoic Era of Earth's history, extending from approximately 145 to 65 Ma ago.

Cyanamide An organic cyanide compound, H_2NCN.

Cyanate An organic radical, NCO^-.

Cyanide An organic radical, CN^-.

Cyanoacetaldehyde An organic cyanide compound, $HO(CH)_2CN$.

Cyanoacetylene An organic compound, HC_3N.

Cyanobacterium Any of a diverse group (Cyanobacteria) of prokaryotic microorganisms capable of oxygen-producing photosynthesis (in older classifications, termed blue-green algae).

Cyanogen A chemical condensing agent, C_2N_2, and the parent chemical radical that gives rise to HCN.

Cyanopolyyne An organic compound, $HC_{11}N$.

Cyclize To undergo cyclization, the formation of one or more rings in a chemical compound.

Cyclohexene An organic compound, the aromatic hydrocarbon C_6H_{10}.

Cysteine A sulfur-containing amino acid, $HSCH_2CH(NH_2)COOH$.

Cystine A sulfur-containing amino acid, $C_6H_{12}N_2O_4S_2$, one of the 20 amino acids commonly present in proteins of living systems.

Cytosine (C) An organic compound ($C_4H_5N_3O$), a pyrimidine, and one of the four nitrogenous bases present in the nucleic acids DNA and RNA. (For chemical structure, see chapter 3 or figure 4.1a.)

Cytosol The watery intracellular cytoplasmic fluid.

D-*Configuration* Pertaining to a stereoisomer of a compound that in pure solution rotates plane polarized light in the *dextro,* rightward direction.

D-*Enantiomer* See D-*isomer.*

D-*Isomer* A stereoisomer of a compound having the D-configuration.

D-*Nucleotide* A nucleotide having the D-configuration.

D-*Ribose* The stereoisomer of ribose sugar present in the nucleic acids RNA and DNA.

Darwinian evolution Biological evolution via natural selection (competition).

Deamination The removal of an amino group, $-NH_2$, from a compound.

Decarboxylation The removal of a carboxyl group, $-COOH$, from a compound.

Dehydration condensation A kind of chemical reaction that involves formation of a water molecule (H–O–H) by removal of an atom of hydrogen (–H) from one monomer and a hydroxyl group (–OH) from another as the two combine into a dimer, or from one monomer and an oligomer that combine into a polymer.

Deoxyribonucleic acid (DNA) The genetic information–containing molecule of cells, a double-stranded nucleic acid made up of nucleotides that contain a nitrogenous base (adenine, guanine, thymine, or cytosine), deoxyribose, and a phosphate group.

Deoxyribose A pentose sugar, $C_5H_{10}O_4$, a structural unit of DNA.

Deuterium The hydrogen isotope ^2H that is twice the mass of ordinary hydrogen ^1H.

Diaminofumaronitrile An organic compound, $C_4H_4N_4$. (For chemical structure, see chapter 3.)

Diaminomaleonitrile An organic compound, $C_4H_4N_4$. (For chemical structure, see chapter 3.)

Diaminopurine An organic compound, $C_5H_6N_6$. (For chemical structure, see figure 3.5.)

Dicarbonyl diimidazole An organic compound having the chemical structure shown in figure 4.15a.

Dicarboxylic acid Any of various organic acids containing two carboxylic acid groups, $(-COOH)_2$.

Dihydroxyacetone An organic compound, $HOCH_2COCH_2OH$.

Dimer A short oligomer consisting of two monomeric subunits.

Dipeptide An oligomer consisting of two amino acids linked by a peptide bond.

DNA See *Deoxyribonucleic acid.*

DNA polymerase A protein enzyme that brings about the replication of DNA.

Domain One of three, major, superkingdom-like primary branches of the Tree of Life.

Dwarf star A stellar object in the final stages of its evolution when its outer layers have been blown away to reveal an underlying hot, much smaller body (such as a white dwarf).

Electron acceptor In a chemical reaction, a molecule that accepts one or more electrons contributed by another molecule.

Electron carrier In a chain of chemical reactions, molecules that accept electrons from an electron donor and pass them to an electron acceptor.

Electron donor In a chemical reaction, a molecule that contributes one or more electrons to another molecule.

$E = mc^2$ Einstein's famous equation for the equivalence of energy and mass.

Electroweak A unified force that combines the electromagnetic and weak nuclear interactions.

Entelechy An inherent essence that regulates and directs the vital processes of an organism that is nonmaterial and not discoverable by scientific investigation.

Enzyme A protein or RNA (ribozyme) capable of catalyzing a biochemical reaction.

Ester Any of various organic compounds containing a carbonyl group (C=O) adjacent to an ether group (C–O–C), as in a compound of the general formal RC(O)COCR in which R is a univalent hydrocarbon radical.

Ethane An organic compound, the hydrocarbon CH_3CH_3.

Ethanol An organic compound, the alcohol CH_3C_2OH.

Ethanolamine An organic compound, $HOCH_2CH_2NH_2$.

Ether Any of various organic compounds characterized by the ether group C–O–C.

Ethylene An organic compound, CH_2CH_2.

Eucarya (eucaryal domain) Together with Bacteria and Archaea, one of three superkingdom-like primary branches of the Tree of Life.

Eukaryote Any of diverse organisms of the superkingdom-like Domain Eucarya, a domain composed of plants and animals—higher organisms having relatively large cells in which chromosomes are packaged in a saclike nucleus.

Europa One of the larger of the 16 satellites of planet Jupiter.

Eutectic Of or relating to an alloy or solution at its lowest melting point.

Exergonic The liberation of energy.

Exobiology Extraterrestrial biology.

Extremophile Any of various microbes, such as many archaeans, that tolerate exceptionally high-temperature acidic environments.

Fatty acid Any of numerous monocarboxylic acids having the general formula $C_nH_{2n+1}COOH$, such as acetic acid, CH_3COOH.

Fatty alcohol Any of numerous alcohols of fatty acids having the general formula $C_nH_{2n+1}CH_2OH$, such as ethyl alcohol, CH_3CH_2OH.

Fermentation Anaerobic metabolism.

Ferric iron Oxidized iron, having a valence of +3 as in the mineral hematite, $Fe_2^{3+}O_3^{2-}$.

Ferrous iron Reduced iron, having a valence of +2 as in the mineral pyrite, $Fe^{2+}S_2^{-}$.

Folic acid An organic acid, $C_{19}H_{19}N_7O_6$.

Formaldehyde An organic compound, CH_2O, the simplest aldehyde and an important intermediate compound in Miller-type organic syntheses.

Formamide An organic compound, $CHONH_2$.

Formamidine An organic compound, $HNCHNH_2$.

Formate A salt of formic acid.

Formic acid An organic acid, $HCOOH$.

Formose reaction See *Butlerov (Butlerow) synthesis.*

Fructose An organic compound, the fruit sugar $C_6H_{12}O_6$.

Furanose A sugar consisting of a five-membered ring like that present normally in the ribose sugar of RNA.

Galaxy A large gravitationally bound and generally spiral or elliptical collection of stars, such as the Milky Way Galaxy, which consists of several hundred billion stars.

Gamma aminobutyric acid (*γ-aminobutyric acid*) An amino acid, $H_2NCH_2CH_2CH_2COOH$, not commonly present in proteins of living systems.

Gas chromatography–mass spectrometry (GC–MS) A technique used to identify organic compounds by the use of a gas chromatographic-mass spectrometric instrument that first volatilizes the compounds then identifies them by their mass spectra.

Gene A segment of DNA containing information for production of one or more molecules of protein or RNA.

Gene copying (gene duplication) The process by which multiple copies of genes are biosynthesized.

Genetic code The biochemical basis of heredity consisting of codons (each a triplet sequence of nucleic acid bases) that determine the specific amino acid sequence in proteins.

Genetic takeover The process in the prebiotic evolution of the Clay World hypothesized by G. Cairns-Smith during which life evolved from a self-replicating inorganic clay into organic-based living systems.

Genetics The genetic makeup of an organism.

Genome The genes present in an organism.

Giant planet Any of the four especially large planets of the Solar System, from innermost to outermost, Jupiter, Saturn, Uranus, Neptune.

Glucose A six-carbon sugar, $C_6H_{12}O_6$, commonly referred to as the universal fuel of life and abbreviated CH_2O.

Glutamic acid An acidic amino acid, $HOOCCH_2CH_2CH(NH_2)COOH$, one of the 20 amino acids commonly present in proteins of living systems.

Glyceraldehyde An organic aldehyde, $CH_2HOCHOHCHO$.

Glycerol An organic alcohol, $CH_2OHCHOHCH_2OH$.

Glycine A chemically neutral amino acid, $CH_2(NH_2)COOH$, one of the 20 amino acids commonly present in proteins of living systems.

Glycoaldehyde An organic aldehyde, $CHOCH_2OH$.

Glycolic acid An organic acid, $HOCH_2COOH$.

Graphite A mineral, crystalline carbon, C_6.

Greenhouse effect A warming of the Earth's surface and lower layers of the atmosphere caused by interaction of solar radiation with atmospheric gases (mainly carbon dioxide, methane, and water vapor) and its conversion to heat.

Greenhouse gas Gases such as ammonia (NH_3), methane (CH_4), carbon dioxide (CO_2), and water vapor (H_2O), that can produce a greenhouse effect.

Guanine (G) An organic compound ($C_5H_5N_5O$), a purine, and one of the four nitrogenous bases present in the nucleic acids DNA and RNA. (For chemical structure, see figures 3.5 or 4.1a.)

Guanylic acid An organic acid, guanosine monophosphate, $C_{10}H_{14}N_5O_8P$.

Half-life The time required for half of the material in a substance to be converted into another substance.

Hematite A mineral, the iron oxide Fe_2O_3.

Hemoglobin An iron-containing protein present in the red blood cells of many animals.

Heredity The transmission of genetic factors that determine individual characteristics from one generation to the next.

Heterooligomer A small compound (oligomer) composed of chemically different parts.

Heterochiral See *Nonracemic*.

Heterotrophic hypothesis The concept introduced by A. I. Oparin and J. B. S. Bernal that the earliest forms of life were heterotrophs that used nonbiologically produced organic matter as their carbon source.

Heterotrophy The metabolic process of animals and animal-like organisms in which organic compounds serve as the source of carbon and energy.

Hexamer A polymer composed of six monomers.

Hexose Any of various six-carbon sugars.

Histidine An amino acid, $C_6H_9N_3O_2$, one of the 20 amino acids commonly present in proteins of living systems.

Homochiral Pertaining to a mixture of molecules that contains only one stereoisomer.

Hubble Space Telescope A robotic telescope, launched by NASA in 1990 by the Shuttle *Discovery*, especially effective because it orbits Earth above much of the atmosphere.

Hydrocarbon Any of a diverse group of organic compounds composed of hydrogen and carbon.

Hydrogen cyanide An organic compound, HCN.

Hydrogen sulfide A gaseous organic compound, H_2S.

Hydrolysis A process of breakdown of chemical compounds that involves addition of the chemical elements of water (HOH) to the substance broken down.

Hydroxy acid An organic acid having the general formula RCH(OH)COOH, in which R is a univalent hydrocarbon radical.

Hydroxy nitrile A cyanide organic acid having the general formula RCH(OH)COOH, in which R is a univalent hydrocarbon radical.

Hydroxyapatite A mineral, $Ca_5(PO_4)_3(OH)$, the material that constitutes bones in vertebrate animals.

Hydroxyl group The two-atom chemical group –OH.

Hydrothermal vent A fissure, commonly on the ocean floor but also on continental platforms, through which hot water, dissolved minerals, and gases emanate.

Hyperthermophile Any of various prokaryotic microorganisms, such as diverse members of the Archaea, that survive and grow at exceptionally high temperatures (>80°C) and generally grow optimally at and above 90°C.

Hypoxanthine An organic compound, a purine having the formula $C_5H_4N_4O$. (For chemical structure, see figure 3.6.)

IDP See *Interplanetary dust particle.*

Illite A clay mineral having the composition $Al,K,Ca,Mg(Si_2O_7)_2(OH)_4$.

Imidazole An organic compound, $C_3H_4N_2$.

Inflationary model A theory of cosmology in which a large cosmological constant temporarily exists early in the history of the Universe, leading to its rapidly accelerating expansion.

Infrared light That part of the electromagnetic spectrum lying outside the visible range at its red end.

Infrared Space Observatory (ISO) Launched by NASA in 1995, an Earth-orbiting telescope that observes in the infrared part of the electromagnetic spectrum.

Infrared spectroscopy A technique used to identify organic compounds by use of their spectral response to infrared light.

Infrared spectrum The distinctive spectrum given by a compound in the infrared part of the electromagnetic spectrum.

Intermediate compound Any chemical compound formed in a reaction sequence prior to formation of the final product.

Interplanetary dust particle (IDP) Any of the small cometary rocky particles (≤ 50 μm) and similarly-sized asteroidal fragments that comprise interplanetary debris.

Interstellar cloud Vast regions of space occupied by gases and tiny particles of matter.

Interstellar dust See *Interstellar cloud.*

Interstellar medium See *Interstellar cloud.*

Ion An electrically charged atom or group of atoms.

Iridium A platinum-group heavy-metal element, ^{77}Ir.

Iron carbide Any of various iron–carbon compounds.

Isocyanic acid An organic compound, CHNO.

Isotope Any of two or more types of atoms of a chemical element that have nearly identical chemical behavior but differ in atomic mass and physical properties.

Isotopic date Age of a rock (or organic substance < 60,000 years old) determined by measurement of the ratio of a stable isotope to one of the products of its radioactive decay.

Isotopic fractionation Separation of isotopes of an element; in organisms, often mediated by an enzyme.

Kaolin A clay mineral having the composition $Al_2(Si_2O_5)(OH)_4$.

Ketone Any of various organic compounds characterized by the presence of a carbonyl group, C=O, attached to two carbon atoms, as in acetone, CH_3COCH_3.

Kinetic energy Energy generated by or resulting from movement.

Kinetic isotopic fractionation Separation of isotopes of an element as a result of their speeds of movement; in organisms, mediated by an enzyme that interacts more readily with one of two or more isotopes of an element.

Kuiper belt A belt of cometary bodies having orbits that lie beyond that of Neptune and mostly beyond that of Pluto.

L-*Configuration* Pertaining to a stereoisomer of a compound that in pure solution rotates plane-polarized light in the *levo*, leftward direction.

L-*Enantiomer* See *l-isomer*.

L-*Isomer* A stereoisomer of a compound having the L-configuration.

L-*Nucleotide* A nucleotide having its isomeric components in the L-configuration.

L-*Ribose* A nonbiological form of ribose sugar, the mirror image of the stereoisomer of ribose present in the nucleic acids RNA and DNA.

Lactic acid An organic acid, $C_3H_6O_3$, produced commonly by fermentation.

Last Common Ancestor (LCA) The plexus of primitive early-evolved microbes that existed before the rise of superkingdom-like domains; the evolutionary rootstock of all organisms living today.

Leucine A chemically neutral amino acid, $(CH_3)_2CHCH_2CH(NH_2)COOH$, one of the 20 amino acids commonly present in proteins of living systems.

Ligate To bond or link together.

Light-year A unit of length equal to the distance that light travels in one year in a vacuum, about 10 trillion km (6 trillion miles).

Limestone A kind of sedimentary rock consisting mainly of calcium carbonate minerals.

Lipid Fats, waxes, and similar organic compounds soluble in nonpolar solvents such as chloroform, $CHCl_3$, and ethyl ether, $C_2H_5CH_2OCH_3$.

Lipoic acid A microbiological growth factor, the organic compound $C_8H_{14}O_2S_2$.

Lysine An amino acid, $NH_2(CH_2)_4CH(NH_2)COOH$, one of the 20 amino acids commonly present in proteins of living systems.

Ma *Mega anna*, one million (1×10^6) years.

Magma Molten rock such as volcanic lava.

Magnetite A mineral, the iron oxide Fe_3O_4.

Main sequence star The phase of stellar evolution, lasting about 90% of a star's life, during which the star fuses hydrogen to helium in its core.

Mandelic acid An organic acid, $C_6H_5CH(OH)COOH$.

Mass spectrometry An instrument-based method of identifying the chemical constituents of a substance by means of the separation of gaseous ions according to their differing mass and charge.

Mesophile An organism adapted to the average range of temperatures present today on Earth.

Messenger RNA (mRNA) RNA molecules that transport genetic information from DNA to ribosomes.

Metabolic pathway Any of numerous enzyme-mediated multistep processes by which metabolism is carried out in living systems.

Metabolism The sum of the energy-consuming and energy-producing processes that happen during the chemical buildup and breakdown of organic compounds in living systems.

Metamorphism A change in the constitution of a rock produced by pressure and heat.

Meteor Solar System matter observable when it falls through Earth's atmosphere and is heated by friction to temporary incandescence; a "shooting star."

Meteorite A meteor that reaches Earth's surface.

Methane The colorless gaseous hydrocarbon CH_4.

Methane-generating prokaryote Any of various archaeal microbes that give off methane gas as a product of their chemoautotrophic metabolism.

Methanogen See *Methane-generating prokaryote.*

Methyl group The four-atom chemical group $-CH_3$.

Methionine A sulfur-containing amino acid, $CH_3SCH_2CH_2CH(NH_2)COOH$, one of the 20 amino acids commonly present in proteins of living systems.

Micelle A molecular aggregate that constitutes a colloidal particle.

Milky Way Galaxy The galaxy of which the Solar System is a part.

Molecular Biologist's Dream A hypothetical world ideal for prebiotic syntheses on the primitive Earth envisioned by G. F. Joyce and L. E. Orgel in which pools and lagoons were loaded with chemically activated nucleotides ready to be polymerized into RNA.

Molecular biology A branch of biology dealing with the molecular physicochemical organization of living matter.

Mollusk Any of an animal phylum (Mollusca) characterized by a large muscular foot and a mantle that secretes spicules or shells, such as a snail, clam, or squid.

Monomer A chemical compound, usually small, that can be linked to other monomers into a larger, multicomponent polymer.

Mononucleotide A nucleotide composed of one monomer each of a purine or pyrimidine nitrogenous base, a ribose or deoxyribose sugar, and a phosphoric acid group.

Montmorillonite A hydrated clay mineral having the composition $(Al, Mg)_8$ $(Si_4O_{10})_3(OH)_{10} \cdot 10H_2O$.

mRNA See *Messenger RNA.*

Murchison meteorite An organic-rich meteorite observed to fall on September 28, 1969 near Murchison, Australia.

Mutation Any change in the nucleotide sequence of a gene.

N-Carboxyanhydride An organic compound (see figure 4.16), a member of a group of similar compounds known as Leuchs' anhydrides.

Natural selection The preferential survival of individuals having advantageous variations relative to other members of their population or species; for natural selection to operate, there must be competition for resources (a struggle for survival) and suitable variation among individuals.

Nebula A region of space containing a higher than average density of interstellar gas and dust.

Negentropic process A process that increases order in a system.

Neurotransmitter A substance that transmits nerve impulses across a synapse.

Neutral atmosphere An atmosphere that is neither reducing (hydrogen-rich) nor oxidizing (such as one containing abundant O_2), e.g., one having the composition $CO_2 + N_2 + H_2O$.

Neutron star A star characterized by an abundance of neutrons, uncharged massive particles present in the nuclei of atoms.

Nitrate A biologically usable form of nitrogen, NO_3^-.

Nitride Any of various nitrogen–metal compounds.

Nitrile An organic cyanide compound that contains the chemical group CN which, on hydrolysis, yields an acid with elimination of ammonia.

Nitrogen fixation The biologic process carried out by various bacterial and archaeal prokaryotes in which molecular nitrogen, N_2, is combined with hydrogen to produce ammonia, NH_3, a biologically usable form of nitrogen.

Noble (inert) gas Any of a group of gases that exhibit great stability and extremely low reaction rates, such as helium, neon, argon, and krypton.

Nonchiral Pertaining to a substance or solution that lacks chirality, or "handedness."

Nonenzymatic Pertaining to a process that takes place in the absence of catalyzing enzyme.

Nonracemic Pertaining to a mixture that contains only one of the two (L or D) stereoisomers of a compound.

Nonpolar Pertaining to an uncharged compound.

Nuclear fusion The fusion of the nuclei of atoms that takes place in stars during some nucelosynthetic processes.

Nucleic acid The genetic information–containing organic acids DNA and RNA.

Nucleoside A chemical compound consisting of a purine or pyrimidine nitrogenous base combined with deoxyribose or ribose sugar.

Nucleosynthesis Any of various processes in stars that result in formation of chemical elements.

Nucleotide Any of several compounds that consist of a sugar (deoxyribose or ribose) linked to a purine (adenine [A] or guanine [G]) or a pyrimidine (thymine [T], cytosine [C], or uracil [U]) nitrogenous base, and a phosphate group; the basic structural units of DNA and RNA.

Nucleus In eukaryotes, a saclike, membrane-enclosed organelle that contains the chromosomes.

Obligate aerobe An organism unable to live in the absence of the molecular oxygen (O_2).

Obligate anaerobe An organism incapable of growth and reproduction in the presence of molecular oxygen (O_2).

Oligomer Any small polymeric chemical compound composed of a few or several monomeric subunits.

Oligonucleotide A short polymer composed of nucleotide subunits.

Oort cloud A spheroidal shell of cometary bodies that encircles the Solar System and is composed of objects having orbits that are from about 100 to more than 10,000 AU distant from Earth.

Organic acid Any of various acidic organic compounds such as fatty acids and carboxylic acids.

Organic chemistry Chemistry that deals with organic compounds, including (but not limited to) chemical compounds and processes that occur in organisms.

Organic compound A chemical compound of the type typical of (but not restricted to) living systems, composed commonly of carbon, hydrogen, oxygen, nitrogen, sulfur, and/or phosphorus.

Orotic acid An organic acid, $C_5H_4N_2O_4$, a biosynthetic precursor of pyrimidines.

Oscillatoriaceae A taxonomic family of morphologically simple filamentous cyanobacteria.

Oxalic acid An organic acid, $(COOH)_2 \cdot 2H_2O$, present in many plants.

Oxic Pertaining to the presence of molecular oxygen (O_2).

Oxidation A chemical process in which oxygen combines with another element (or in which hydrogen or electrons are removed from an element) and energy is released in the form of heat.

Oxide A chemical compound consisting of oxygen and one or more other elements as in the minerals hematite, Fe_2O_3, and magnetite, Fe_3O_4.

Oxygenic photosynthesis Oxygen-producing photoautotrophy such as that in plants, algae, some protists, and cyanobacteria.

Ozone A triatomic form of oxygen, O_3, formed naturally in the upper atmosphere.

p-RNA See *Pyranosyl-ribonucleic acid.*

PAH See *Polycyclic aromatic hydrocarbon.*

Paleobiology A broad, paleontology- and geochemistry-based science that, by seeking evidence of the living processes of ancient life, goes beyond the traditional fossil-focused approach.

Panspermia The theory that living microbes can be transported from one star system to others by the radiation pressure of starlight.

Paraformaldehyde An organic compound, polymerized formaldehyde, $(CH_2O)_n$.

Pentose Any five-carbon sugar.

Peptide A chemical compound, such as a protein, in which components are linked by peptide bonds (see figure 4.2b).

Peptide bond The type of chemical bond that links adjacent amino acids in proteins.

Peptide nucleic acid (PNA) A nucleic acid analog in which the sugar–phosphate backbone of RNA or DNA is replaced by a backbone held together by amide bonds rather than the standard sugar–phosphate linkage (see figure 5.6c).

Petrify The process by which organic-walled organisms can be preserved as fossils, embedded three-dimensionally in a siliceous, calcitic, or other mineral matrix.

pH A scale used in expressing acidity and alkalinity, the values of which extend from 0 to 14; pH = 7 represents neutrality, pH < 7 indicates acidity, and pH > 7 indicates basicity.

Phanerozoic Eon The younger of two principal divisions (eons) of Earth's history, extending from the beginning of the Cambrian Period, ~550 Ma ago to the present; the Phanerozoic and the older Precambrian Eon comprise all geologic time.

Phenylacetaldehyde An organic compound, the aldehyde $C_6H_5CH_2CHO$.

Phenylacetylene An organic compound, the hydrocarbon C_8H_6.

Phenylalanine An aromatic amino acid, $C_6H_5CH_2CH(NH_2)COOH$, one of the 20 amino acids commonly present in proteins of living systems.

Phosphate ester bond A particular kind of chemical bond present in all nucleic acids.

Phosphate group The chemical group, PO_4^{3-}.

Phosphodiester bond See *Phosphate ester bond.*

Phospholipid Any of various phosphate-containing lipids.

Phosphorus nitrile An organic cyanide, $P(CN)_5$.

Phosphorylation The addition to a compound of phosphorus or of a phosphorous-containing chemical group.

Photoautotrophy Autotrophy powered by light energy.

Photodissociation The breaking apart of a chemical compound by photolysis.

Photolysis Decomposition of a chemical compound by the action of light energy.

Photosynthesis The metabolic process carried out by photosynthetic bacteria, cyanobacteria, algae, and plants in which light energy is converted to chemical energy and stored in molecules of biosynthesized carbohydrates (e.g., in oxygenic photosynthesis, light energy $+ CO_2 + H_2O \rightarrow CH_2O + O_2$).

Photosynthetic Bacterium Any of diverse types of bacteria capable of anoxygenic photosynthesis.

Physiology Pertaining to the chemical functions and activities of living systems.

Planetesimal Any of various celestial bodies, meters to tens of kilometers across, that aggregate to form planets by gravitational attraction.

Plate Tectonics Global tectonics based on an Earth model characterized by many thick oceanic or continental plates that move slowly across the global surface propelled by movement of underlying material of the planetary interior.

PNA See *Peptide nucleic acid.*

Polar Pertaining to compounds having dipoles, positively and/or negatively charged parts.

Polycyclic Aromatic Hydrocarbon (PAH) Any of various organic compounds composed of a few to many usually six-membered rings of carbon atoms linked by an alternating sequence of single and double bonds and to which hydrogen atoms are attached.

Polymer A multicomponent chemical compound, usually large, that consists of many monomers linked together.

Polymerase An enzyme that builds polymers from monomeric subunits.

Polynucleotide Any of various nucleic acids, polymers of nucleotides.

Polypeptide Any of various proteins, polymers of amino acids.

Polysaccharide Any of various carbohydrates, polymers of monosaccharides.

Porphyrin Any of various metal-free derivatives of pyrrole, characteristic especially of chlorophyll or hemoglobin.

Pre-protein Any of various kinds of molecules hypothesized to have preceded and evolved into proteins.

Pre-RNA Any of various kinds of molecules hypothesized to have preceded and evolved into RNA.

Precambrian Eon The older of two principal divisions (eons) of Earth's history, extending from the formation of the planet, 4,550 Ma ago, to the beginning of the Cambrian Period, ~550 Ma ago; the Precambrian and the younger Phanerozoic Eon comprise all geologic time.

Prokaryote Any of diverse types of non-nucleated microorganisms of the Archaea and Bacteria.

Propionic acid An organic acid, CH_3CH_2COOH.

Protein A polymeric organic compound composed of amino acid monomers.

Proteinoid Any of various organic polymers, in some respects similar to proteins, formed by the heating of mixtures of amino acids.

Proto-Earth The early Earth, in the process of forming.

Protoplanet A planet that is in the process of forming.

Proto-Sun The early Sun, in the process of forming.

Protobiont A precellular or very primitively cellular living system hypothesized to have preceded the emergence of cellular life.

Protogalaxy A galaxy that is in the process of forming.

Proton–proton chain A sequence of nucleosynthetic reactions that transforms hydrogen to helium.

Protoplanetary disk A discoidal concentration of gas and dust that can condense to form a planet by gravitational attraction.

Purine A type of nitrogenous base in nucleic acid, such as adenine or guanine.

Pyranosyl-ribonucleic acid (p-RNA) A ribonucleic acid in which the five-membered furanose form of ribose is replaced by its six-membered isomer, the pyranose form (see figure 5.6b).

Pyranose A sugar consisting of a six-membered ring; for the sugar ribose, a form not normally present in RNA.

Pyridoxal An organic compound, $C_8H_9NO_3$.

Pyrimidine A type of nitrogenous base, such as cytosine, thymine, or uracil, present in nucleic acids.

Pyrite A mineral, iron sulfide, FeS_2.

Pyrolysis A breakdown of compounds due to intense heating.

Pyrophosphate A chemical form of phosphate like that in the triphosphates present in living systems.

Pyrrole An organic compound, C_4H_5N, arranged in a ring consisting of four carbon atoms and one nitrogen atom.

R group An abbreviation meaning "chemical radical" used in generalized formulas of organic compounds to represent a univalent hydrocarbon radical such as the methyl group, $-CH_3$, or the ethyl group, $-C_2H_5$.

Racemic mixture One in which equal amounts of a compound are present both in the stereoisomeric configuration that occurs in biologic systems (such as the L form of amino acids and the D form of sugars) and in the nonbiological mirror image of this configuration; for example, equal mixtures of L- and D-isomers of amino acids. (For a comparison of the amino acid L-alanine and a mixture of L- and D-alanine, see figure 4.13.)

Racemize A process during which compounds that are present initially in a single isomeric configuration become changed such that the result contains equal amounts of compounds both in their original configuration and in the mirror image of this configuration.

Radioactivity The property possessed by isotopes of some elements (such as carbon and uranium) of spontaneously emitting alpha or beta rays by the disintegration of the nuclei of atoms.

Red giant A phase in stellar evolution following completion of the main-sequence phase, when the star becomes especially large and bright.

Reducing atmosphere An atmosphere that is rich in hydrogen and hydrogen-containing gases having, for example, a composition of $CH_4 + N_2 + H_2O$, $CH_4 + NH_3 + H_2O$, $CO + N_2 + H_2O$, or $CO_2 + N_2 + H_2$.

Reduction In chemistry, to combine with or subject to the action of hydrogen, thereby lowering the oxidation state.

Ribonucleic acid (RNA) A normally single-stranded nucleic acid made up of nucleotides that contain a backbone of ribose sugar and phosphate to which are linked the purines (adenine and guanine) and the pyrimidines (cytosine and uracil).

Ribonucleotide A nucleotide containing ribose.

Ribose A five-carbon sugar, $C_5H_{10}O_5$.

Ribose phosphate A dimer consisting of ribose and a phosphate group.

Ribosomal ribonucleic acid (rRNA) Any of several RNAs present in ribosomes.

Ribosome The tiny globular "protein factories" where protein synthesis occurs in cells; made up of three RNA molecules and scores of proteins.

Ribozyme Any of numerous kinds of RNA that have enzymelike properties.

RNA See *Ribonucleic acid.*

RNA World A stage in prebiotic, precellular evolution hypothesized by W. Gilbert during which life originated as self-replicating ribozyme-like RNAs that served both as information-containing and catalytic molecules.

rRNA See *Ribosomal ribonucleic acid.*

Sarcosine An organic acid, CH_3NHCH_2COOH.

Sediment Particulate, commonly granular mineralic material deposited by water, wind, or glaciers that can be lithified to a sedimentary rock on compression.

Seed plant Any of diverse higher land plants, such as gymnosperms and angiosperms, that produce seeds.

Semipermeable Pertaining to a barrier, such as a membrane, that is permeable to small molecules but not to other, usually larger, particles.

Serine A hydroxyl-containing amino acid, $HOCH_2CH(NH_2)COOH$, one of the 20 amino acids commonly present in proteins of living systems.

SETI Abbreviation for the Search for Extra-Terrestrial Intelligence, a project begun by NASA in 1992 and currently carried out by the privately funded SETI Institute in northern California.

Shale A sedimentary rock formed by consolidation of clay or mud.

Shoemaker-Levy 9 (SL-9) A comet that was observed to fall onto the upper layers of the atmosphere of planet Jupiter in 1994.

Silica An inorganic chemical compound, silicon dioxide (SiO_2), the mineral of which sand is composed.

Silicate Any of a great number of silicon- and oxygen-containing minerals.

Sodium hydroxide An inorganic chemical base, NaOH.

Solar mass The mass of the Sun, 1.99×10^{33} g, about 330,000 times the mass of Earth.

Solar nebula The rotating circumstellar disk and its embedded protostar that gravitationally condensed to form the Solar System.

Solar System The Sun and the planets (and their satellites) that orbit it, from innermost to outermost: Mercury, Venus, Earth, Mars (the terrestrial planets); Jupiter, Saturn, Uranus, Neptune (the giant or Jovian planets); and Pluto (a planet-like object).

Solar (stellar) wind The continuous or quasicontinuous ejection of plasma from the Sun's surface into and through interplanetary space.

Species In biology, the fundamental category of biological classification, ranking below the genus and in some species composed of subspecies or varieties.

Spectrum The electromagnetic radiation given off by an object or substance, usually plotted as a function of wavelength or frequency.

Spontaneous generation The theory that life could originate all at once from dead matter, a concept disproved by the French chemist Louis Pasteur in the 1860s.

Spore plant Any of various lower land plants such as club mosses (lycophytes) and horse-tails (sphenophytes) that reproduce by shedding spores rather than by producing seeds.

Stereoisomer Any of a group of isomers in which atoms are linked in the same order but differ in their spatial arrangement.

Stratum A layer or horizon of sedimentary rock.

Strecker synthesis A chemical reaction sequence by which amino acids are formed from aldehyde, hydrogen cyanide, and ammonium.

Succinic acid An organic acid, $HOOCCH_2CH_2COOH$.

Sugar Any of various monosaccharides having the generalized formula CH_2O, such as fructose, sucrose, and glucose.

Sulfide A chemical compound of sulfur and another element as in the mineral pyrite, FeS_2, or gaseous hydrogen sulfide, H_2S.

Sulfuric acid An inorganic acid, H_2SO_4.

Supernova The violent explosion of a star at the end of its life that generates heavy elements extending from iron to uranium.

Tectonic Pertaining to a crust-deforming process or event.

Template-directed synthesis A chemical synthesis in which a polymeric molecule serves as a substrate (a template) that organizes and stabilizes the formation of another polymer on its surface.

Terrestrial planet Any of the four innermost rocky planets of the Solar System, from innermost to outermost: Mercury, Venus, Earth, and Mars.

Tetrose Any of various four-carbon sugars.

Thermodynamics Physics that deals with the mechanical action or relations of heat.

Thermophile Any of various organisms, such as diverse prokaryotes, that can survive and grow in relatively high-temperature environments such as hot springs.

Thiamine An organic compound, vitamin B_1, $C_{12}H_{17}N_4OS)Cl$.

Thioacetic acid A sulfur-containing organic acid, C_2H_4OS.

Thioester World A stage in prebiotic, precellular, evolution hypothesized by C. De Duve during which chemical reactions among thioester-containing organic compounds (sulfur-linked ester compounds) played a central role.

Tholin A term coined by C. Sagan to refer to the complex organic polymer formed in a Miller-type prebiotic synthesis.

Threonine A hydroxyl-containing amino acid, $CH_3CH(OH)CH(NH_2)$ COOH, one of the 20 amino acids commonly present in proteins of living systems.

Thymine (T) An organic compound, $C_5H_6N_2O_2$, a pyrimidine and one of the four nitrogenous bases present in the nucleic acid DNA.

Titan One of the larger of the 16 satellites of planet Jupiter.

Toluene An organic compound, C_7H_8, known also as methylbenzene.

Transcribe In molecular biology, the process by which the sequence of bases in DNA is encoded into messenger RNA (mRNA) by RNA polymerases; a process similar to that involved in DNA replication.

Translate In molecular biology, the complicated process by which the sequence of bases in messenger RNA (mRNA) is used to produce functional proteins at ribosomes.

Tree of Life A branching, treelike representation showing the relatedness of all living organisms, commonly based on comparison of rRNAs, the ribonucleic acids of protein-manufacturing ribosomes.

Tricarboxylic Acid Any of various organic acids containing three carboxylic acid groups $(-COOH)_3$.

Trilobite Any of a group (Trilobita) of extinct arthropods of the first half of the Phanerozoic Eon (the Paleozoic Era, 543 to 245 Ma ago), characterized by a three-lobed body organization.

Triple-α process A process of nucleosynthesis by which one atom of ^{12}C is formed by the condensation of three helium nuclides.

Tyrosine An aromatic amino acid, $C_9H_{11}NO_3$, one of the 20 amino acids commonly present in proteins of living systems.

Ultraviolet (UV) light That part of the electromagnetic spectrum having wavelengths between 200 and 300 nm that lie outside the visible range at its violet end.

Universal Tree of Life A branching, treelike representation showing the relatedness of all living organisms, commonly based on comparison of rRNAs, the ribonucleic acids of protein-manufacturing ribosomes.

Uracil (U) An organic compound, $C_4H_4N_2O_2$, a pyrimidine and one of the four nitrogenous bases present in the nucleic acid RNA. (For chemical structure, see figure 4.1a.)

Uranyl ion A charged uranium oxide radical, UO_2^{2+}.

Urea An organic compound, $CO(NH_2)_2$.

Valine A chemically neutral amino acid, $(CH_3)_2CHCH(NH_2)COOH$, one of the 20 amino acids commonly present in proteins of living systems.

Virus A submicroscopic organic object, typically composed of a protein coat surrounding an RNA or DNA core of genetic material.

Visible light That part of the electromagnetic spectrum visible to humans and lying between the infrared and ultraviolet parts of the spectrum.

Weathering The chemical and physical processes that disaggregate a rock into its component mineral grains or crystals.

White dwarf Evolutionary end-point of stars that have initial masses less than about eight times the solar mass.

Xanthine An organic compound, the purine $C_5H_4N_4O_2$. (For chemical structure, see chapter 3.)

Index

Compositor: TechBooks
Text: 10/13 Sabon
Display: Sabon
Printer and Binder: Edwards Brothers, Inc.